Galileo Galilei

A Captivating Guide to an Italian Astronomer

(That's How the Man Who Changed the Path of Science)

Stanley Reaves

Published By **Simon Dough**

Stanley Reaves

All Rights Reserved

Galileo Galilei: A Captivating Guide to an Italian Astronomer (That's How the Man Who Changed the Path of Science)

ISBN 978-1-77485-805-9

No part of this guidebook shall be reproduced in any form without permission in writing from the publisher except in the case of brief quotations embodied in critical articles or reviews.

Legal & Disclaimer

The information contained in this ebook is not designed to replace or take the place of any form of medicine or professional medical advice. The information in this ebook has been provided for educational & entertainment purposes only.

The information contained in this book has been compiled from sources deemed reliable, and it is accurate to the best of the Author's knowledge; however, the Author cannot guarantee its accuracy and validity and cannot be held liable for any errors or omissions. Changes are periodically made to this book. You must consult your doctor or get professional medical advice before using any of the suggested remedies, techniques, or information in this book.

Upon using the information contained in this book, you agree to hold harmless the Author from and against any damages, costs, and expenses, including any legal fees potentially resulting from the application of any of the information provided by this guide. This disclaimer applies to any damages or injury caused by the use and application, whether directly or indirectly, of any advice or information presented, whether for breach of contract, tort, negligence, personal injury, criminal intent, or under any other cause of action.

You agree to accept all risks of using the information presented inside this book. You need to consult a professional medical practitioner in order to ensure you are both able and healthy enough to participate in this program.

Table Of Contents

Chapter 1: A Stargazer Is Born 1

Chapter 2: Galileo Studies With Florentine Monks 8

Chapter 3: The University Of Pisa ...15

Chapter 4: Galileo Calculates The Location Of Hell 21

Chapter 5: Professor At The University Of Pisa 33

Chapter 6: University Of Padua 40

Chapter 7: The Catholic Inquisition 48

Chapter 8: Kepler's Star 55

Chapter 9: Galileo And Johannes Kepler 63

Chapter 10: The Starry Messenger .72

Chapter 11: Galileo Meets Pope Paul V 82

Chapter 12: The Inquisition Visits Again 89

Chapter 13: Discourse On The Tides 100

Chapter 14: A Meeting With Pope Urban Viii 107

Chapter 15: The Assayer 118

Chapter 16: Dialogue Concerning The Two Chief World Systems 127

Chapter 17: Trial And Imprisonment 136

Chapter 1: A Stargazer Is Born

1564 was an ominous year for Europe, a continent that was in the midst of the Scientific Revolution. Future astronomer Galileo Galilei was born on February 15 of that year, and just three days later, the great artist and natural philosopher Michelangelo di Lodovico Buonarroti Simoni—better known only by his first name—died in Rome.[1] That same April, William Shakespeare was born in Stratford-upon-Avon in England.[2] Europe was running full tilt into a new age of science, art, and religious change that would become known as the Age of Enlightenment.

As it had during the Renaissance, Italy once again led the charge in creating a world full of art, education, and philosophy. One of those Italians was

Vincenzo Galilei, an accomplished composer, musician theorist, and lute player. Born in Florence, Vincenzo and his wife, Giulia (née Ammannati), settled in Pisa to raise their family, and their first child was Galileo. Partial recordkeeping means that historians must make a few assumptions about the names of Galileo's siblings, but they are usually believed to have been named Virginia, Michelagnolo (or Michelangelo), Livia, Giulia, and Benedetto Galilei. Galileo had five siblings in total, but only three survived past infancy.

As was natural in a musical household, the elder Galilei taught his children how to play his own preferred instrument, the lute. He did not necessarily have musical aspirations for his offspring, but his youngest son, Michelagnolo, would grow up to become a musician. Though Vincenzo's firstborn son would not pursue music as a career, Galileo did become a skilled lute player and musical theorist under the close tutelage of his father. In fact, Vincenzo and Galileo worked together on much of Vincenzo's musical experimentation, and Galileo was an important part of his father's study of acoustics.

Together, they studied the mathematics of music, focusing on the rules governing the sounds created by the strings of their lutes. In the basement of their home in Pisa, Galileo and Vincenzo scrupulously stretched various lengths of lute strings (at the time usually made from dried sheep's intestines) across the room and affixed different weights to the ends. They used different lengths and weights to measure the variations in sound across all of the strings. The pair's experiments succeeded in uncovering a set of mathematical constants at work with acoustic sound theory. By watching his father work methodically at the string experiments, young Galileo learned a great deal about scientific methodology that would not even be generally accepted for centuries to come.

At some point after the birth of his children, Vincenzo Galilei found it necessary to move to the nearby city of Florence for his work. His family remained behind in Pisa in the household of Vincenzo's friend, Muzio Tedaldi. Tedaldi saw to the care and education of the Galilei children in Vincenzo's absence, during which time young Galileo most likely attended Pisa's public school. As a student of the Renaissance, Galileo would have studied a classical-style collection of subjects, including Latin, algebra, philosophy, history, music, and European languages. These subjects had faded out of popularity during

Europe's Dark Ages, but the educated classes of the Renaissance were dedicated to placing them once more at the forefront of education.

While Galileo and his siblings spent their early childhoods in Pisa, Vincenzo remained in Florence as he was a musician in great demand. True to the era of the Renaissance, Galileo's father focused on a Greek musical revival that included ancient musical theory as well as dramatic plays. He convened with the noted Florentine Camerata (also known as the Camerata de' Bardi), a group comprised of musicians, poets, philosophers, and other highly educated men under the patronage of Count Giovanni de' Bardi. Together with the Camerata, Vincenzo helped transform the burgeoning art form that was Italian opera.

Specifically, Vincenzo was fixated on the musical idea of "dissonance," which refers to the perceived darker, harsher parts of a musical piece. He explored the use of this type of music in dramatic productions, theorizing that there was a use for it within larger compositions. Vincenzo was also largely responsible for the use of recitative lyrical forms in opera, where singers would mimic normal speech patterns instead of traditional song form. The recitative style became an almost fundamental part of many operatic forms, which still continue to be used to this day.

Vincenzo's firstborn son would have the honor of becoming known as one of the world's greatest minds, and just the great Florentine Renaissance Man who came before him—Michelangelo—he would also be recognized primarily for his first name only. Galileo Galilei would transform the fields of astronomy and physics as much as his father transformed Italian opera, but the former would suffer more for his work.

Chapter 2: Galileo Studies with Florentine Monks

Florence, Italy, was the epicenter of the European Renaissance as early as the 13th century, but despite its cultural influence over the rest of the continent, Florence itself was home to a great deal of political and religious strife. Under the strict control of the rich and powerful de' Medici family, the city prospered in terms of the arts but grew more and more corrupt in terms of politics and taxation.

Florence was inextricably linked to the de' Medicis for over three centuries, during which the family gained power over the city via their banking monopoly. In the 14th century, Giovanni de' Medici invented a new form of public taxation and banking that gripped the popular market city of Florence and would not let go. Giovanni used much of his money to create new, beautiful buildings and patronize artists like the famous Donatello. His son, Cosimo di Giovanni de' Medici, inherited his father's banking wealth and used it to solidify his political position in Florence. A learned patron of the arts, Cosimo also used much of his money to fund artworks, libraries, and learning centers within the city.

After Cosimo, however, the de' Medici family heirs became less inclined to patronize the arts and culture of their city, though no less than three of them were granted the highest office in the Catholic Church, that of the pope. Their banking empire crashed by the late 15th century, and public sentiment in Florence toward the family grew resentful; the city's residents pushed the de' Medicis out twice in the span of a century, creating a republic. Both times, however, the de' Medicis' hired armies forced their way back into power, and in 1532, Pope Clement VII—born Giulio di Giuliano de' Medici—disbanded the government of the republic and established Alessandro de' Medici the Duke of Florence. Henceforth, the city-state operated under the official authority of the de' Medici line of dukes. In 1569, Pope Pius V declared Cosimo I de' Medici to be the Grand Duke of Tuscany, incorporating the Florentine city-state into the greater Grand Duchy of Tuscany.

When young Galileo Galilei was sent for by his father to take up residence in Florence in 1574, the city was still a cultural and scientific hub despite its ongoing political unrest. Sixteenth-century arts and culture had persisted throughout Florence's chaotic recent history, and by the time the Galilei family moved to the city from Pisa, Francesco I de' Medici had just inherited the position of Grand Duke of Tuscany. Perhaps due to a mixture of traits inherited from his father, Cosimo I, Francesco allegedly ordered the murder of his mistress' husband and simultaneously founded the Florentine porcelain and stoneware industries.

For the Galilei family, the disturbing behavior of the Grand Duke of Tuscany was probably of little consequence as Francesco gladly supported the musical and dramatic arts. During his time as the leader of the Grand Duchy of Tuscany, Francesco had the Medici Theatre and the Accademia della Crusca built. Vincenzo had plenty of work, and Galileo was surrounded by all the sciences of his day: architecture, music, philosophy, philology, mathematics, and natural philosophy. The boy stayed with his father in the city until 1579, which was when he was sent to study with local monks.[3]

The fifteen-year-old pupil may have studied at the monastery of Vallombrosa, or he may have been educated in their beliefs and customs in a nearby community such as Santa Trinita. Both locations were within the Grand Duchy of Tuscany, as Cosimo de' Medici had managed to purchase a great deal of the city's outlying lands. Regardless of the exact location of his study, Galileo found great personal belief in the Catholic religion taught to him by the monks. He enjoyed his time with the Vallombrosans and possibly even formally entered their order as a young man. His devotion to the religion he had been taught there would continue throughout his life. Of course, as was customary during the 16th century in Europe, studying with a religious order did not mean one would only be educated in religious doctrine. What was known of science, mathematics, and philosophy were also taught at such institutions, and Galileo's education would not have been an exception.

There are notes written in Galileo's own handwriting that date to this period which show his study of Aristotle's logic. Unfortunately, Galileo's father did not want his son to complete his education with the monks because that would tie him to a career with the Church. Instead, the elder Galilei wanted his son to study medicine and become a practicing physician. The Vallombrosan abbot Diego Franchi seems to have implied that Galileo's father made an excuse "of taking him to Florence to treat a severe eye condition."[4] Evidently, the student did not return. Galileo was taken from the Vallombrosan order in 1578, and two years later, he was enrolled at the University of Pisa as a student of medicine and philosophy.[5] Once more, he was situated at the house of Muzio Tedaldi.

Chapter 3: The University of Pisa

After ending his education with the Vallombrosan monks, Galileo followed through with his father's wishes and looked ahead to an education at a university. He was able to attend the University of Pisa, one of the most well-respected educational institutions in all of Europe at the time.

Though the university had originally been established in Florence in 1343 as a studium generale, it was relocated to Pisa by the de' Medicis in 1545.[6] Studium generale is the medieval term for what would eventually be formally known as a university. These institutions welcomed people from all over the world who wished to gather together and study any subject. Classes were not as strictly scheduled and formalized as they would be in the following centuries, though by the time Galileo Galilei attended the

University of Pisa, modern-style degree programs had been created.

At that point, Pisa had been politically joined with its long-time rival Florence, thanks to it having been sold to the Duchy for 206,000 florins.[7] Under the authority of the de' Medici family, the two cities became closely intertwined, and Duke Cosimo I de' Medici reopened the University of Pisa officially in 1545.[8] His patronage greatly improved the standing of the school, which became one of the most prestigious in Europe. With his home city under the political umbrella of the Grand Duchy of Tuscany, Galileo had the best education in the region within reach.

The teenaged Galileo did as he was expected and attended medical studies at the university, learning the more complex mathematics and biological topics equated with a contemporary medical degree. Instead of becoming interested in medicine and patient care, however, young Galileo found himself much more fascinated by the scientific study of objects and their movements. Not unlike his work with Vincenzo and the lute strings, Galileo was attracted to the as-yet unnamed science of physics. He wondered about the motions of the stars, the Earth, and the things upon its surface.

Though he continued to attend the school under the pretense of earning his medical degree, Galileo actually spent most of his time at the University of Pisa taking mathematics and natural philosophy classes. Mathematics was one of the most fascinating topics of the day for students interested in higher learning, and much of the materials studied by Galileo would have been directly influenced by the works of an Italian friar named Luca Pacioli. Pacioli published several mathematical books whose contents included geometric puzzles, the Fibonacci sequence, and double-entry bookkeeping.

Pacioli died in 1517, but his books were highly regarded within Italy and abroad, and these would have made their way to Galileo's own desk several decades later.[9] Pacioli is possibly the first mathematician to have used positive and negative symbols in his work, which had a great impact on the field. Throughout the century, various mathematicians began to incorporate new symbols into the lexicon, including those for multiplication, division, equality, fractions, roots, and decimals. Galileo probably found great enjoyment in working through Pacioli's puzzles, some of which may have been designed or even drawn by his friend and colleague, Leonardo da Vinci.

It was probably through works like Summa de arithmetica, geometria, proportioni et proportionalita (Summary of arithmetic, geometry, proportions and proportionality) that Galileo would have learned about geometry, basic algebra, and accountancy—all the important tenets of 16th-century mathematics. Most of these fundamentals were directly applied to astronomy, an ancient but relatively simple study that involved mapping the sky, charting the movement of heavenly bodies, and completing equations to assist with navigation. Most navigational mathematics used trigonometry, which Galileo and his contemporaries studied from the texts of the Greeks. All of these relied on the use of the naked eye and rudimentary tools.

As it was, Galileo was not adequately interested in medicine to continue with his studies. Just as he had done at the Vallombrosan school, Galileo quit before earning a degree. Thus, both the careers of clergyman and physician were out of reach; however, the young man was by no means unemployable. He was highly educated, intelligent, and familiar with many subjects. Therefore, he made an ideal teacher. Finished—for the moment—with Pisa, the 21-year-old Galileo Galilei made his way to Florence to begin work as a mathematics teacher.

Chapter 4: Galileo Calculates the Location of Hell

Happily, the young professor found his talents in demand at his favorite academy, the Vallombrosan monastery. He worked there during the summer of 1586, probably thrilled to have found a career in which he could be surrounded by the comforts of his faith while also being given leave to explore the world by way of numbers and equations. It was during this time that he began to put together his first publication: La Bilancetta (The Little Balance).

La Bilancetta was a contemporary explanation of an ancient Greek mathematical process. In the book, Galileo described the methods Archimedes used to find specific gravities of objects using a balance. "Specific gravity" is the ratio of an object's density to the density of a different substance.

For example, one might test the specific gravities of 100 grams of lead and 100 grams of wood by checking them against the same material—probably water.

Archimedes is famous for having shouted "Eureka!" after getting into a bath, as he had finally figured out how to measure oddly shaped objects without melting them down. He realized, during the process of getting into his bath and watching the water level rise, that he could measure the mass of an object by placing it into a bath and noting how much water was displaced. The Archimedes' principle states that the buoyant force on an object is equal to the amount of water it displaces.

Galileo's Bilancetta examined Archimedes' principle, as well as Archimedes' law of the lever, a law in which the ratio of objects' weights can be determined. The experiment uses a long lever set upon a balance point and a water bath. Two identical objects are suspended from opposite ends of the lever, with one of the objects sitting in the water. The difference between the weight of the objects (with one suspended in the air and the other made lighter by the buoyant forces of the water) allows one to determine the object's density.

As a man of numbers, Galileo's reputation grew ever greater. In 1588, he was invited to lecture on the subject of the specific location and dimensions of Hell as portrayed in the famous Dante's Inferno. For a culture of people whose daily lives revolved around the doctrine of the Catholic Church, theology concerning the contents of the Bible was considered to be of the utmost importance. Even men of science (generally speaking) believed in the basic truth of Catholicism, and, of course, Galileo was no exception.

Durante di Alighiero degli Alighieri, more commonly known as Dante Alighieri, was a 14th-century Italian poet famous for writing the Divine Comedy, an epic poem comprised of cantiche, in which the first is entitled Inferno. The second and third cantiche are entitled Purgatorio and Paradiso. Inferno follows the path of Dante, led by ancient Roman poet Virgil, through the nine levels of Hell as he has envisioned them. Dante's model of Hell is located inside of the Earth, with each level descending deeper into the core of the planet. These descriptions were based on Dante's own research of the Bible and several other theological texts, including those of the Muslim world. In Dante's vision of Hell, the most heinous of sinners were brought to the innermost chamber of the Earth while lesser sinners dwelled in the outer layers.

The medieval clergy took this particular theology very seriously, and when Galileo was asked to work with Dante's poem and use his advanced knowledge of mathematics to determine the exact size and location of the levels of Hell, he took the project seriously as well. As he was only 24 years old, Galileo must have been a very prestigious citizen of Florence to have been considered for the job. His own location probably had a great deal to do with it. At that point in his career, Galileo was a respected teacher, a published mathematical author, and a faithful member of the Church. Residing in Florence meant that he was closer to the ultimate seat of power—the de' Medicis and the pope—than similar great minds in other parts of Europe.

The materials that Galileo had to work with were all based on Dante's original cantica; however, several artists and theologians had come up with their own physical representations of Hell. Sandro Botticelli, the artist who had created a map of Hell for a version of the Divine Comedy, was probably more influential to this process than Dante himself since his artwork had inspired other men to consider the physical realities of the spiritual underworld. A few decades later, in 1506, the Florentine mathematician and architect Antonio Manetti wrote his own description of the geography and geometry of Dante's Inferno.[10] This was considered to be a remarkable feat until it was challenged in 1544 by Alessandro Vellutello.[11] Alessandro posited quite a different map from that of Manetti, and thus was born the need for Galileo to provide a well-formed critique of both descriptions.

The task was not without political pressure, however. Since Vellutello was from Lucca, an Italian city with which Florence had an ongoing rivalry, the assembly was biased toward the Florentine scholar Manetti. A Florentine himself, and quite young, Galileo may have been biased himself before even reading through Vellutello's work. Indeed, after delivering two greatly detailed lectures to the Florentine Academy on his findings, Galileo was in favor of Manetti's description of Hell. For his part, however, the mathematics professor clearly attempted to make a clear-cut case for his decision.

The following text is an excerpt from one of Galileo's lectures on the subject of the location of Hell.

The size and depth of the Inferno is as great as the radius of the earth, and its mouth, which is the circle turned about Jerusalem, has for its diameter an equal size, because under the arc of the sixth part of the circle is a chord equal to the radius. But wanting to know its size in respect to the whole volume of earth and water, we should not just follow the opinion of some who have written about the Inferno, who believe it to occupy the sixth part of the volume, because making the computation according to the methods proved by Archimedes in his book On the Sphere and the Cylinder, we will find that the space of the Inferno occupies a little less than 1/14 part of the whole volume; I say this if that space should extend all the way to the surface of the earth, which it doesn't: on the contrary, the mouth remains covered by a great vault of earth, whose summit is

Jerusalem and whose thickness is the eighth part of the radius, which is 405 [and] 15/22 miles...

[The] levels go turning round and round the concavity of the Inferno; and the first, the nearest to the surface of the earth, is Limbo; the second is that where the sensuous are punished; in the third are castigated the gluttonous; the fourth holds the prodigal and the avaricious. The fifth level is divided into two circles, the first of which includes the Stygian swamp and the moats around the city, the place assigned to the pains of the wrathful and the sullen; the second the city of Dis, where the heretics are punished.

And here one should point out that by levels we do not mean what are called circles by Dante, because we propose that the levels are distinct from each other by greater or lesser distance from the center, which isn't always the case with circles, witness that in the fifth level the Poet places on the same level two circles. But because the other levels are still called circles by the Poet, we can say that in all there are 9 circles and 8 levels. Next therefore follows the sixth level and seventh circle, the torment of the violent, which is divided into 3 rings, so named by the Author.

And here we can note the distinction which Dante makes between circles and rings, the rings being parts of the circles, like this seventh one, divided into 3 rings of which the one surrounds the other. And the first and the greatest in circumference, which is a lake of blood, surrounds the next, which is a forest of stumps, which surrounds the third ring, which is a plain of sand...

It seems to me that these arguments can persuade us how much more plausible the Inferno of Manetti is than that of Vellutello.[12]

Galileo calculated that the immense roof of Hell not only encompassed Jerusalem but that it also stretched as wide to the west as Marseilles, France, and as far to the east as Tashkent, Uzbekistan. In order not to collapse upon itself, the mathematician believed that the roof must be 600 kilometers (373 miles) thick. This second calculation was determined using the famous duomo from the Florence Cathedral as a reference, as it was a marvel of contemporary architecture. A duomo is a massive domed roof, and this particular duomo sits upon the highest tower of the Florence Cathedral. Built by Filippo Brunelleschi in the first part of the 15th century, the famous dome measures between thirty and sixty centimeters (between one and two feet) thick.

Ironically, Galileo soon realized that his calculations were incorrect when he discovered the true relationship between the dimensions of a roof and its volume. By that time, however, his work with the Florence Assembly had solidified his brand-new job at the University of Pisa. He would not publicly address the error he had made until five decades had passed.

Chapter 5: Professor at the University of Pisa

It was in an atmosphere of triumph that Galileo returned to the university from which he had never graduated, though, of course, the new professor was carrying the weight of his unknown mistake with him. The year after his meeting with the Florence Academy, the scientist was given the prestigious position of the Chair of Mathematics at the university where he had become a professor.[13] The job wasn't as well paid as one might assume, but the fame and opportunity to explore mathematics on a daily basis was all Galileo had dreamed of.

Just as the committee appointed to the Florence Academy had been, the University of Pisa was very much under the authority and leadership of the Catholic Church. The texts used by the students and teachers—including any that Galileo and his colleagues wrote themselves—had to be approved by appointed overseers. The university was considerably strait-laced, with all scientific endeavors conforming to the approval of the pope and his advisors. Forced to keep his secret concerning the mistake he had made in calculating Dante's vision of the nine layers of Hell, Galileo's lesson plans focused on other mathematical and physical concepts.

Galileo was interested in the way the forces of the universe played out upon the Earth, including the not yet named force of gravity. In accepting that the known planets moved around the Sun, Galileo and other scientists also postulated that there was an attracting force between the planets and the Sun— an attracting force that could also explain why loose objects fell to the ground. In a famous set of experiments observed by mathematics students at the University of Pisa, Galileo determined to discover the effect of the Earth's attractive force on objects of different masses.[14]

To find evidence of any differences between the speeds at which different items fell to the ground, Galileo designed an experiment that involved dropping orbs of different masses from the top of the Leaning Tower of Pisa and recording which reached the ground the fastest.[15] It was from this series of experiments that our modern classrooms took their lesson of the feather and the hammer, both of which fall at the same speed in the absence of friction.

During his tenure at the University of Pisa, Galileo amassed enough data on the rules of physical motion to write his next book, a series of essays entitled De

Motu Antiquiora (The Older Writings on Motion). The book references information that he had gained during experiments conducted with students, including those where he had thrown things off the Leaning Tower, which was compared to known laws concerning physics. In particular, he compared the results of his motion experiments with the ideas of Aristotle. In fact, Galileo had found that orbs of different masses fell at the same speed, an observation that negated Aristotle's historical assertion that objects fall at a speed relative to their mass. His law of falling bodies states that free-falling bodies are in a constant state of acceleration, and furthermore, the distance a falling body travels is proportionate to the square of the elapsed time of the fall.

De Motu Antiquiora was never published, though Galileo copied his chapters out several times, revising them and eventually switching literary forms from an essay to a dialogue. The latter parts may have been written later in his life, as this was the same literary form he used to write one of his last books. Regardless of when each chapter was written, the collection wouldn't be published until decades after his death. Nevertheless, it was obviously an important book to Galileo, who kept it with him until he died.

Perhaps Galileo worried that he would make the same mistake as before and be forced to retract the contents of the book. Whatever the reason he chose not to publish his findings on falling objects, Galileo was sure to impress upon his students the importance of using the proper scientific method. He explained to them that it was necessary to design concise experiments that could be repeated by anyone to achieve the same results. Whole centuries before the formal theory of the scientific method was accepted by the professional community, Galileo and a few of his contemporaries knew its value. By experimentation, he believed, one could test theories to prove or disprove them.

It is difficult to say exactly why Galileo decided to leave his prestigious post at one of the most highly accredited universities in Italy, but it may have been that he felt stifled by the omnipresence of the Catholic Church. When Galileo tired of the University of Pisa, he was fortunately not starved for other options. Therefore, he did not need to look far to find another academic post. The young man was lucky not only in having been born to a family of some means and importance but also in having been born in the cradle of Europe's expanding fields of arts and education. Another academic institution, 285 kilometers (177 miles) away, was happy to welcome the famous Galileo Galilei, professor of mathematics, to their halls: The University of Padua.

Chapter 6: University of Padua

Located in northern mainland Italy, just across the sea from Venice, the University of Padua welcomed many world-famous scientists through its doors during the Renaissance and the Age of Enlightenment. Already having been one of Italy's premier post-secondary institutions for over two centuries before Galileo stepped foot in its halls in 1592, the northern university was not only prestigious, but it was quite a bit more liberal for the age.[16] As they were located much closer to neighboring European countries, the cities of Padua and nearby Venice were frequented by many international students, natural philosophers, and even non-Catholic theologians.

Established in the 13th century and overseen by the Republic of Venice, the University of Padua had never been under the supervision of the Catholic regime. Even in the late 16th century, at the peak of the Catholic Inquisition, Padua was far enough from Rome and the Vatican that its lessons were not censored by the pope. Nicolaus Copernicus, one of the most eminent astronomers of the early 16th century, had attended lessons at the school and received his medical doctorate there. The university was home to the world's first anatomical theater, which was located in one of the university buildings. The university also boasted the world's first botanical garden, which had been built in 1545.[17]

Furthermore, Galileo received three times the salary for teaching at Padua than he'd received at Pisa. Thus began, in Galileo's own words, "the happiest time of my life."[18] Padua, like the University of Pisa, had an academic program mostly geared toward students studying for a medical degree. Still, Galileo was kept busy teaching those students the basic tenets of higher mathematics, which mostly concerned Euclid's geometry and standard astronomical calculations.

At the time, astronomy was an important factor in the medical profession since patient horoscopes would advise physicians as to the unseen maladies affecting their clients. Such horoscopes were drawn up for the time period in which the patient had first become ill. Then, depending on the patient's date of birth and zodiac classification, the physician would find the recommended treatment. The practice was heavily interconnected with the classical belief in the body's four humors, which physicians attempted to balance in order to set one's health back to normal. Due to its close association with the medical field,

Galileo probably discussed horoscopes on a daily basis with his students and colleagues and believed in them himself.

In fact, once he had moved to Padua, Galileo began taking on private clients and calculating their horoscopes to make extra money outside the salary he earned as a professor. The horoscope of an individual in the 16th century was considered to be very important, not only because of its believed relevance to medical matters but also because of its perceived ability to forecast the future. A mathematician's and an astronomer's ability to draw up one's horoscope was seen as an incredibly valuable—and frightening—skill. It was actually illegal to draw up the horoscope of a member of the royal family unless instructed to do so directly by the monarch; searching the stars for the date of a king's or queen's death was strictly forbidden.

The other main part of Galileo's lesson plans were the thirteen texts of the ancient Greek mathematician, Euclid. Euclid's books were called Elements, and they were primary texts used in Renaissance Europe. Elements focuses largely on geometry but also contains theorems and proofs related to prime numbers, algorithms, and trigonometry. There was plenty of content in Euclid's texts for Galileo to delve into, but, of course, his growing passion was for astronomy.

In an era when the Greek philosophers were considered to be at the forefront of their studies, Aristotle's views on astronomy were the most commonly taught. Galileo, however, found his long-dead predecessor's theories on physics and heavenly bodies rife with misunderstandings. Whereas Aristotle had taught his own students that the matter comprising Earth and the heavens were different, Galileo believed they had more in common than not. Galileo was also a nonbeliever in Aristotle's theory that like matter was attracted to other like matter—such as how smoke was attracted to the sky because it was more like air than earth.

It wasn't just academic freedom and the ability to question the Greeks that Galileo found in Padua and Venice but social freedom as well. He began a romantic relationship with Marina Gamba, a woman who bore him three illegitimate children. For his time, Galileo was by no means alone in having such a relationship, and though it would have been frowned upon for Marina to make social calls alongside Galileo (given her apparent low social ranking), the professor wouldn't have been considered a bad Catholic by most contemporaries. His three children by Marina were born in 1600, 1601, and 1606, and they were named Virginia, Livia, and Vincenzo, respectively.

None of the children's birth records, however, list Galileo as the father, though he was and is still considered to be the only candidate. Marina may have even been living in Galileo's house in Padua, which would have been an unusual arrangement between unmarried lovers back then. Nevertheless, Virginia's and Livia's birth records are blank where the name of the father is usually written, and on Vincenzo's, it says "father uncertain." Galileo still cared for his children and his lover as well as if they were all legitimate, even reaching out for extra income via horoscope writing after the births of his children.

Unfortunately, with fame and esteem came the closer, critical eye of the Catholic Church. Less than a decade after Galileo made the move from Pisa to Padua, he began to face dangerous allegations from members of the clergy concerning those very horoscopes that were at the heart of his income.

Chapter 7: The Catholic Inquisition

No matter where one lived or worked in 16th-century Europe, the Catholic Inquisition was never too far away. The Inquisition was a gathering of organizations inside the Catholic Church whose point was to battle blasphemy. The Inquisition began in 12th-century France to battle great religious differences in the population, and the group's special inquisitors were usually picked from the Dominican Order of the Catholic Church. Before and during the Renaissance, the idea and extent of the Inquisition fundamentally extended in light of the Protestant Reformation, coming full circle in the provincial Spanish Inquisition and Portuguese Inquisition. These extended all the way into parts of Africa, Asia, and the Americas.

Though it was true that Europeans who lived farther from Rome or Spain were more likely to be left alone by the Catholic Church, no place was completely safe. So it was that even in the liberal city of Padua, Galileo had critics who believed his methods of creating horoscopes were heretical. On April 22, 1604, a formal charge was filed by the Catholic Inquisition against Galileo Galilei. He was accused of "haver ragionato che le stelle, i pianeti at gl'influssi celesti necessitino." In essence, the charge meant that Galileo was accused of practicing deterministic astronomy, which was forbidden by the Church.

The accusation seems to have come from an employee in Galileo's household, a man named Silvestro Pagnoni. Working for Galileo as an assistant, Silvestro complained to the local office of the

Inquisition that his boss was on poor terms with his own mother for keeping a mistress and three bastard children in his house. Silvestro also claimed that Galileo was not attending mandatory Catholic mass and that he was performing deterministic astronomy for his rich clients.

Deterministic astronomy was founded on the idea that everything that came to pass on Earth had already been determined when the universe began. To practice such a form of astronomy—at least in terms of horoscopes—meant that the astronomer believed all things were metaphorically written in the stars. To an extent, the Church condoned finding answers in the movement of the stars, but to accept that all things were predetermined was a thing the clergy could not do. The very idea contradicted the part of the Bible that claimed God had given his creatures free will to do as they decided.

Galileo, despite his persistent Catholicism, believed that information could be found in the stars, and therefore, certain truths of the world and life upon it had been predetermined. Silvestro himself was questioned by the Inquisition, and though it seems clear that he had a great dislike for his employer, he stated that he was unaware of any specific heresies committed by Galileo. Though the scientist was personally interrogated over the accusations, ultimately, the charge was not sought after by higher courts, and it never went to the Holy Office in Rome. Luckily for Galileo, the Church did not wish to begin an issue with the school at Padua. This whole affair was actually overlooked for centuries until the first summons was found by Franciscan Friar and Professor Antonino Poppi. Poppi found these records in the Sartori documents at Padua in 1990, incorporating two sworn revilements in the Venetian State Archives.[19]

Of course, it is unsurprising that Galileo would not wish to advertise his entanglement with the Inquisition, even if his case was dropped. Once the embarrassing inquiries came to an end, the professor refocused his energies on work that took place quite apart from horoscopes. 1604 was a busy year for him; he invented a water-lifting machine for use at the Contarini estate in Padua, experimented with falling bodies on inclined planes, and visited the duke of Mantua in search of more lucrative employment.

Vincenzo Gonzaga was the duke of Mantua at the time, and he was, without a doubt, a standard benefactor of technical studies and the arts. Truth be told, his spending had put the once-booming city in monetary trouble. Having been mentored by Giuseppe Moletti, Galileo's forerunner as the Chair of Mathematics at the University of Padua, Mantua was glad to offer the candidate a position in his court. Galileo was offered compensation of 300 ducats for each year at Mantua's court in addition to everyday costs for himself and an assistant. His pay at the University of Padua was already 320 ducats, however, and he was also paid from outside jobs and student boarders. Attempting to negotiate a better deal, Galileo asked for compensation of 500 ducats with a business account for himself and assistants. No agreement could be found, so Galileo remained at Padua.

That December, he observed a new phenomenon in the sky that would greatly influence his work in the upcoming year.[20]

Chapter 8: Kepler's Star

Galileo was not the first astronomer in Europe to view the supernova of 1604; in fact, he wasn't even the first in Padua to do so. Indeed, the modern concept of a supernova had yet to be invented, since Galileo and his contemporaries knew nothing of the inevitable explosions of the universe's stars. The first scientist in the continent to observe the supernova and document its appearance was Lodovico delle Colombe of Florence, though the astral body would come to be known as Kepler's Star. Johannes Kepler, an eminent astronomer of Germany, began watching the exploded star and tracking its journey in early October of 1604, and it was because of his detailed study that the phenomenon got its name.

The new star could be seen by the naked eye, and contemporary records of its sighting exist in Chinese, Arabic, and Korean sources. Kepler's Star glowed more brightly than any other star during the first few weeks of its discovery, even glowing visibly throughout the daytime for more than fourteen days. The mysterious shining object caused a great sensation among the scientific community in Padua and Venice, and Galileo was invited to present a series of lectures on the subject.

In those lectures, which took place in the Padua University hall, Galileo told his audiences that the new star was small at first but that it grew quickly in size until it was bigger than all the other stars in the sky. With the exception of Venus, it was even bigger than the planets in our solar system. The new star sparkled, seeming red when at its darkest and golden at its brightest. Galileo said, "Someone could therefore reasonably conjecture that it was generated by the embrace of Mars and Jupiter, the more so that it seemed to be born at the time of their encounter, which took place in the presence of Saturn at 5:00 p.m. on 9 October."[21]

The lectures were a great success, but there was controversy among the scientists who had helped Galileo form his conclusions. In making his grand presentations day after day to a packed hall, the astronomer had failed to give credit to his colleagues, namely, Baldassarre Capra. Capra had actually been present the night of October 4 to observe the emergence of the new star, and it was by collaborating with Capra that Galileo had been able to fill in many of the details for his lectures.

Capra's claims continued after the series of lectures came to a close, and evidently, Galileo made no real attempt to apologize or even give his colleague a share of the credit for discovering the new star. The next year, Capra published an essay in which he explained how Galileo could not have possibly seen the great conjunction of the planets the previous October because, in Padua, the sky had been overcast. Galileo responded in kind, publishing his own treatise on the incident. Galileo's own claim was that he had properly thanked Capra and other colleagues on the evening of the first lecture, and that furthermore, it was less important who saw the star first than who made the most correct assumptions about its physical properties.

To Galileo, the interpretation of the data was the primary goal. He surmised—using the data collected from Capra and other scientists throughout Europe—that the new star was more distant than the

Moon. To illustrate his point, he used the concept of parallax. Astronomers estimate the exact length of neighboring objects in the distance using a procedure referred to as stellar parallax. In other words, they measure a star's apparent movement from the backdrop of distant stars since Earth revolves around sunlight. A good example of parallax is how the location of an object seems to move when you close one eye.

To say that Kepler's Star was more distant than the Moon meant that it belonged in the realm of the distant, tiny, seemingly unmoving stars beyond our solar system. Johannes Kepler agreed, which meant that both famous scientists were directly challenging the age-old belief of Aristotle that the far distant stars were unchanging and fixed forever in place. If Aristotle was correct, there was no place for any new objects at such a distance from the Earth. Galileo puzzled at how such a star—obviously quite different from normal stars—could have come into being.

One of his theories was that such a huge amount of vapor had escaped from the Earth and collected together in the distant cosmos. He posited that huge clouds of smoke could rise into the sky from a wood fire without diminishing the size of the wood; therefore, perhaps the Earth could give off such vapors without losing a great deal of its own mass. Eager to find the answer, Galileo consulted the work of Tycho Brahe, Kepler's recently deceased employer and the primary astronomer behind the amazing island laboratory Uraniborg. Brahe had been the first to record a similar event (also a supernova) in 1572.[22] Brahe's work was illuminating, but Galileo hesitated to draw any solid conclusions about

Kepler's Star. As always, he carried the knowledge of his mistaken claims about Hell's roof, and he was extremely cautious about making any such mistakes again.

In 1606, delle Colombe printed Discourse of Lodovico delle Colombe, in which he posited that the star was neither a passing comet nor was it in fact "new." Delle Colombe defended an Aristotelian opinion of cosmology that Galileo had already challenged in his lectures and in his own writing, suggesting that the star had been present in the sky all along; it had just been simply too dim to see. This was, ironically, quite true, as it was only by means of exploding that Kepler's Star became bright enough to observe in the sky. The competing Aristotelian and Copernican camps continued to speculate, and in doing so, Galileo and

Johannes Kepler found much to discuss via their frequently exchanged letters.

Chapter 9: Galileo and Johannes Kepler

Galileo Galilei was unquestionably one of the most eminent scientists of the day, but he was far from the only person in Europe searching the skies, making elaborate calculations, and fine-tuning the tools of his profession. In fact, between the 16th and 19th centuries, hundreds of amazing minds contemplated the questions of their world and even formed clubs by which they could keep track of one another's experiments and theories.

Despite being separated by a multitude of states and tiny principalities, scientists of similar minds forged bonds during the Renaissance. For Galileo Galilei, perhaps one of the most important individuals in his network was Johannes Kepler, a man with whom Galileo maintained a regular correspondence. Kepler was not just the scientist responsible for compiling the most sought-after collection of data concerning the 1604 supernova, but he had also worked with Denmark's most famous astronomer and scientist, Tycho Brahe, at the latter's royal-funded science castle on an isolated island off the coast of Denmark. Called Uraniborg, Brahe's decadent observatory and adjacent laboratories were a major attraction for Europe's prominent scientists.

Kepler became an astronomical assistant to Tycho Brahe in the late years of the 16th century, and the two worked together to make detailed, constant observations of the movement of the Sun, Moon, stars, and planets. Other scientists visited Uraniborg regularly to check out the immense wealth of data Brahe and his assistants had compiled, as well as to convene together with their peers. Kepler fought to combine his beliefs with those of his employer, but despite a difference in opinion, working as Brahe's assistant brought him closer to other established members of the industry.

After Brahe's death in 1601, Kepler was finally able to examine every piece of natural philosophy and astronomical data that had been monopolized by Brahe, including much of Copernicus' work. This was a great boon to Kepler's own studies, which had been in support of the Copernican model of a heliocentric universe; it has even been speculated that Kepler poisoned his employer to gain access to this treasure trove of documents, but there is no substantial evidence to prove this to be true. His old colleague, Brahe, had not subscribed to the same model, instead hypothesizing that the Sun orbited the Earth but that all other planets orbited the Sun. Finally out from under the shadow of Tycho Brahe, Johannes Kepler could finally see his own ideas gain some traction in Europe's scientific communities.

In Kepler's mind, the Sun lay in the middle of everything. It was not an entirely new means of looking at things, as the Greek 3rd-century BCE philosopher Aristarchus had written the same idea long before. But even in Galileo and Kepler's time, it failed to catch on. Kepler was frustrated that the political climate didn't allow him to publish his theories without the fear of violent backlash, a feeling that Galileo knew all too well. The two astronomers corresponded regularly on the subject, with Kepler urging his friend to look outside of Italy for a place to publish his theories without any risk of being investigated for heresy. If only Galileo would express his ideas on the page, his friend insisted, Kepler would lend his own name to the venture. The latter hoped that together they would be able to turn the tide of astronomy and change the way the contemporary world thought about the universe.

[Letter from Galileo to Kepler, 1597]

Like you, I accepted the Copernican position several years ago and discovered from thence the causes of many natural effects which are doubtless inexplicable by the current theories. I have written up many of my reasons and refutations on the subject, but I have not dared until now to bring them into the open, being warned by the fortunes of Copernicus himself, our master, who procured immortal fame among a few but stepped down among the great crowd (for the foolish are numerous), only to be derided and dishonored. I would dare publish my thoughts if there were many like you; but, since there are not, I shall forebear.[23]

[Kepler to Galileo, 1597]

I could only have wished that you, who have so profound an insight, would choose another way. You advise us, by your personal example, and in discreetly veiled fashion, to retreat before the general ignorance and not to expose ourselves or heedlessly to oppose the violent attacks of the mob of scholars (and in this you follow Plato and Pythagoras, our true preceptors). But after a tremendous task has been begun in our time, first by Copernicus and then by many very learned mathematicians, and when the assertion that the Earth moves can no longer be considered something new, would it not be much better to pull the wagon to its goal by our joint efforts, now that we have got it under way, and gradually, with powerful voices, to shout down the common herd, which really does not weigh the arguments very carefully? Thus perhaps by cleverness we may bring it to a knowledge of the truth. With your arguments you would at the same time help your comrades who endure so many unjust judgments, for they would obtain either comfort from your agreement or protection from your influential position. It is not only your Italians who cannot believe that they move if they do not feel it, but we in Germany also do not by any means endear ourselves with this idea. Yet there are ways by which we protect ourselves against these difficulties...Be of good cheer, Galileo, and come out publicly. If I judge correctly, there are only a few of the distinguished mathematicians of Europe who would part company with us, so great is the power of truth. If Italy seems a less favorable place for your publication, and if you look for difficulties there, perhaps Germany will allow us this freedom. [24]

Kepler painstakingly examined existing Copernican data, as well as new data, and cross-referenced it with the Copernican theory. The data seemed to hold, just as Galileo had also observed in Italy, but the latter would not publish a major book on the subject until 1610. Kepler, on the other hand, was a bit more brash about his scientific theories, probably because he lived in the part of the Holy Roman Empire that would become Germany, a location in which Protestantism ruled. Protestants were initially less concerned with the potentially controversial works of scientists than the Catholic Church since their primary troubles had to do with theology. Before either of these scientists dared to speak openly about their support of the Copernican model, they decided to write about their work with a brand-new invention: the telescope.

Chapter 10: The Starry Messenger

Several people, including Galileo himself, have been credited with the invention of the refracting telescope in 1607.[25] Most scholars agree, however, that the true inventor was Hans Lippershey. Hans was a Dutch eye-glass manufacturer who applied for a patent for the telescope, which was described as a tool "for seeing things far away as if they were nearby."[26] Though Lippershey is evidently the first optician credited with the application of a patent for what is clearly a basic telescope, he was not awarded the patent. The basis of the patent rejection from the Netherlands office was that just a few weeks after Lippershey's application, another Dutch optician, Jacob Metius, applied for a patent on a similar design. Believing the technology to already be in the public domain, the patent office rejected both applications.

Rapidly, the invention did indeed become a fascination for many lens-crafters throughout Europe. Many scientists who had no previous experience grinding or polishing glass lenses took up the hobby so they could work on their own magnification tools—Galileo Galilei included. Galileo's fascination with the original instrument invented by Lippershey led to many modern advancements and uses of the telescope. He documented the changes he made to the original model and wrote about the various ways he used the improved instrument in Siderius Nuncius (read as Sidereal Messenger or as Starry Messenger), a book published in 1610. The book outlined the methods in which Galileo adopted Hans Lippershey's spyglass and modified it to magnify objects up to thirty times their actual size. Galileo also described the objects he looked at through the improved set of magnification lenses, noting the differences he found between reality and common theory concerning their physical nature. The instrument was innovative, but until then, it had only been used to magnify earthbound objects and distances. Galileo had a different idea for his own telescope, however, and so, he pointed his skyward to take in the light of the stars. Apparently the first recorded astronomer to do so, Galileo's observations of the night sky led him to an important conclusion: Earth was not the only planet in the universe.

The things he saw in the sky had never been observed before by any human in history, and they must have been quite shocking. First, he inspected the Moon. Common belief held that the surface of the Moon must be perfectly smooth and tranquil, as were all objects in the sky since they comprised the spaces of Christian Heaven. Through his eyepiece, however, Galileo saw that the Moon was full of rough land and immense craters—exactly the opposite of what he had expected to see.

Using his own telescope, Galileo looked up at the Moon, of which he could only view about one quarter at a time due to the small diameter of the instrument. He was excited and shocked to discover that the surface of the Moon was not smooth, as had been assumed, but that it was covered in pockmarks, hills, and valleys. This was to be the first major astronomical discovery of many. Using his later thirty-magnification telescope, Galileo was also able to observe the planets of our solar system. It was January 7, 1610, when he first laid eyes on three small stars surrounding Jupiter.[27] Though one lay to the west of Jupiter and the others to the east, the following night, Galileo found all three stars to the west of the planet. Soon, he noticed a fourth member of the group, and he realized that he had not found stars but a collection of moons orbiting Jupiter. These were Io, Ganymede, Callisto, and Europa.

By turning his creations toward Venus, Galileo was also able to document the phases of the planet. They were much like those of the Moon, which suggested Venus orbited around the Sun, not the Earth—the latter a theory posited both by Tycho Brahe, Copernicus, and Johannes Kepler. Evidence that the Earth was not the center of the universe was piling up, and it must have been an extremely exciting time for Galileo— albeit simultaneously frightening, as he would have worried about the repercussions of such discoveries. Nevertheless, he published his discoveries and named the Jovian moons "Cosmica Sidera," or "Cosimo's Stars," in honor of Cosimo II de' Medici, the current Grand Duke of Tuscany.

Since Galileo's book proposed that the newly discovered moons of Jupiter ought to be named after Cosimo II de' Medici, the duke's astronomical birth chart was an important addition to the manuscript. It was a grand but logical gesture, as the de' Medicis remained the most powerful patrons in Florence, and the planet Jupiter was associated with royalty. Galileo certainly had no qualms about flattering his powerful friends and patrons, and it was probably this tendency that kept him repeatedly out of the dangerous grasp of the Catholic Inquisition. With the de' Medici family very much on Galileo's side—particularly when, at Cosimo's suggestion, Galileo changed the name of his moons to "Medicea Sidera"—he probably felt much more confident in his continued work proving the Copernican theory of the universe to be correct. Ironically, the names that stuck to these newly found moons were chosen by Simon Marius, a German astronomer who made the same discovery at almost the same time as Galileo.

Without having improved upon Lippershey's refracting telescope, Galileo, Simon Marius, and other astronomers could never have come to such conclusions as the ones presented in Siderius Nuncius. Lippershey's original refracting telescope used two lenses placed at a distance from one another within a long barrel to produce a magnified view, and Galileo copied this method during his first work with lenses. He carried out a number of experiments using various curves and sizes of lenses, and he actually took up the craft of glass shaping in order to create lenses to his own precise requirements.

Painstakingly, Galileo sanded and polished convex and concave lenses in an attempt to learn how they best worked together to sharpen one's line of sight. The original telescope presented an upside-down image to the viewer because of how one convex and another concave curved lens interacted, just like the human eye. In his experiments, Galileo figured out how to use a bi-curved piece of glass to right the final image and make it more user-friendly.

One of the key features of those early telescopes was length, since the farther apart the two lenses were, the higher a magnification was possible. Galileo's first finished telescope measured nearly a meter (three feet) in length and was constructed using wood and leather. The viewing window was just 2 millimeters (.08 inches) across, so the instrument was able to magnify only a very small area at a time.[28]

The telescopes for which Galileo became famous for were constructed with a biconcave lens at the eyepiece and a planoconvex lens within the instrument. Because of the shapes of these delicate pieces of glass, as well as the distance between them, higher magnification was within reach than had been possible with Lippershey's single-sided (encompassing both convex and concave curves) lenses. In just a few years, the telescope would become the primary instrument of astronomers everywhere. Thanks to Galileo's own designs, the data they helped scientists collect was more precise and detailed than ever before.

Chapter 11: Galileo Meets Pope Paul V

Galileo knew very well that to present astronomical findings that directly challenged those of the Catholic Church was dangerous, and yet he could not help his own inquisitive nature. In learning the secrets of the universe, he also learned to act neutral upon the subject of heliocentrism versus geocentrism. He was constantly searching for a patron, however, who would support his true beliefs and allow him to publish without the fear of criticism and religious persecution.

Since Galileo's desire for support centered on the Catholic Church, he decided to confront the church face-to-face. In the early 17th century, that meant meeting with Pope Paul V. Paul V was an uncompromising leader whose primary issue was the connection of church and state affairs. He enforced the rules of his church rigorously, leading to a range of disputes with multiple cities and states. The dispute with the independent city of Venice in 1606 was one of the most serious events of his papacy, nearly resulting in a full-scale European war. Intending to grasp Venice in the powerful grip of the Holy Roman Empire, Paul V was unable to place his own authority higher than that of the city's government, which depressed him greatly.

There were international concerns as well. In 1606, some Catholics in England were put to death, which included the well-known figure Guy Fawkes, for their role in what is known as the Gunpowder Plot. This was of particular concern to the pope since England had been a Protestant country for nearly a century by that point, and the failed Gunpowder Plot not only vilified the distant country's disenfranchised Catholics, but it also meant that the organized revolution of those same Catholics had failed.

The Gunpowder Plot, also called the Jesuit Treason, was an unsuccessful plan by provincial English Catholics to murder the Protestant King James I. The English Catholics involved in the secret organization of the plot were led by Robert Catesby. His arrangement was to explode the House of Lords during the State Opening of Parliament on November 5 as the prelude to further revolutionary actions planned in the English Midlands. During the second part of the revolt, King James' nine-year-old daughter, Elizabeth, was to be introduced as the replacement Catholic head of state. Catesby may have set out on this plan after the attempts of his fellow Catholics to negotiate with King James had failed, leaving numerous

English Catholics disillusioned with the idea that there would be no renewal of Catholicism in the country.

The plot was uncovered to the local authorities via a letter sent to William Parker, 4th Baron of Monteagle, in late October. The letter was anonymous but effective. Parker set events in motion that led to a raid on the House of Lords on the fourth day of November, one day before the Gunpowder Plot was scheduled to have taken place. There, Guy Fawkes was found guarding 36 barrels of explosive black powder, which would have been enough to decimate the House of Lords to rubble. Most of the plot's schemers fled England's capital city when they learned of the capture of Guy Fawkes, but they still attempted to recruit more support for their cause as they ran.

A few undiscovered plot members remained in London, holding fast against the pursuant Sheriff of Worcester and his men, but they didn't last long. During the fight that ensued, Robert Catesby was killed, and eight others were arrested. At their subsequent court date, the captured plotters were found guilty of treason and were condemned to be hanged, drawn, and quartered.

The news understandably concerned Pope Paul V, who publicly declared that the work of the English Catholic revolutionaries was no less than God's work. He, like all Catholic popes since the reign of English King Henry VIII, was frustrated that England refused to turn back to the Church. Believing that only his religion was valid and true, the pope would not condemn the actions of men whose ultimate goal was the reestablishment or protection of the Catholic faith.

The ongoing repercussions of these events continued to weigh heavily on the mind of Pope Paul V when he met Galileo Galilei in 1616. Galileo hoped to use the meeting to champion his belief in the Copernican theory of heliocentrism, which clerics tended to see as heresy. The pope was not enthusiastic about the theory, but he knew that Galileo was a Catholic in good standing, and therefore, he chose not to focus much attention on the subject. Paul V had his attendant, Cardinal Robert Bellarmine, take on the matter so that he could be left to handle more pressing concerns.

Cardinal Bellarmine was instructed to dictate to Galileo that he should not personally advertise his support of the Copernican theory. The topic was not off-limits, Galileo was assured; however, he was expected not to teach it as truth nor tell anyone that Copernicanism was his preferred world model. Via Cardinal Bellarmine, the pope assured Galileo he wouldn't be placed on trial or pursued by the Catholic Inquisition simply for teaching a variety of worldviews—and that was the best Galileo could hope for given the political and religious environment at the time. He requested a hard copy of that promise in writing and received it in the form of a letter written by the cardinal.

Galileo kept that letter safe for the remainder of his life. Pope Paul V hastily returned to the subject he felt was most important: the revolution of Catholicism against Protestantism in all parts of Europe.

Chapter 12: The Inquisition Visits Again

Despite Pope Paul V's condoning Galileo's continued work, there already existed evidence that Galileo did indeed favor the Copernican model of the universe. The popularity of Siderius Nuncius supported the obvious assertion that Galileo not only theorized about a heliocentric universe but that he also believed other theories to be based on nonsense. This was against the exact instructions of the pope and the cardinal, and though they did not become personally involved in any case against the astronomer, Galileo was once again investigated by the Catholic Inquisition.

Almost immediately following Galileo's meeting with the pope, the Inquisition once more found reason to investigate Galileo—and they evidently had already been doing so for some time. This time, the inquisitors were concerned with the scientist's alleged belief that the Earth moved around the Sun. He had unwittingly cultivated many critics in the clergy by pursuing the Copernican theory that the Earth did not lie at the center of the universe, and these people actively accused him of blasphemy. The inquisitors took their time gathering evidence and interviewing witnesses, but the main piece of evidence was a letter that Galileo had sent to fellow mathematician and former student Benedetto Castelli, which was subsequently sent on to the Dowager Grand Duchess Christina of Tuscany.

In the letter, Galileo told Castelli how frustrating it was to have biblical phrases taken so literally by people of the church, remarking that the phrase "the hands of God" was not meant to be translated as fact but figuratively as in God's existence in the lives of his people. Considering the

Bible must perhaps not be translated in most scenarios, Galileo claimed, it's senseless supporting an individual perspective of their world when a single person cannot fathom God's truth. "Who would dare assert that we know all there is to be known?"[29]

These were a series of private letters, but they were published by Galileo in the hopes that the discussion would spark a kind of understanding between religion and science. They were inspired by an occurrence that took place within the de' Medici household in 1613, during which Cosimo II de' Medici and his mother, the Grand Duchess Christina, began discussing Jupiter's satellites during breakfast. Benedetto Castelli, Galileo's student who was also present, asked Galileo to comment on the central point of that conversation. That point was the seeming fallacies of Galileo's proposals concerning the universe when taken in contrast with the specific words of the Bible.

Galileo's reply to Castelli's request was the famous "Letter to Grand Duchess Christina," which was originally sent to Castelli himself. Eventually, most of the contents of that letter were circulated widely in manuscript form. In its pages, Galileo famously declared that the Bible teaches people how to go to Heaven, not how the heavens work. Galileo's belief in the truth of the Copernican hypothesis alarmed Dominican Inquisitors such as Tommaso Caccini and Niccolò Lorini, and their council investigated Galileo's letter to Christina.[30]

The letter was extensive, reaching seven pages in length, and it seems to have served an important purpose to a man whose greatest work had to be at least somewhat concealed from the Spanish Inquisition. Galileo was luckier than his contemporary, the astronomer Giordano Bruno, because at least his beliefs about a heliocentric universe didn't get him burned at the stake—and yet, Bruno's death in 1600 had to have been a major warning sign to the poor man's scientific successors. Still, Galileo insisted that scientific research must continue despite its contrast with biblical texts.

In his own words, Galileo pleaded with the de' Medici duchess to consider another viewpoint:

Though the Scripture cannot err, nevertheless some of its interpreters and expositors can sometimes err in various ways. One of these would be very serious and very frequent, namely to want to limit oneself always to the literal meaning of the words; for there would thus emerge not only various contradictions but also serious heresies and blasphemies, and it would be necessary to attribute to God feet, hands and eyes, as well as bodily and human feelings like anger, regret, hate and sometimes even forgetfulness of things past and ignorance of future ones.

Thus in the Scripture one finds many propositions which look different from the truth if one goes by the literal meaning of the words, but which are expressed in this manner to accommodate the incapacity of common people; likewise, for the few who deserve to be separated from the masses, it is necessary that wise interpreters produce their true meaning and indicate the particular reasons why they have been expressed by means of such words.[31]

Given the scientific principles in which Galileo believed, he had to be constantly on guard to defend his findings from theologians. Having already written his letter to Benedetto Castelli two years earlier, the scientist found it necessary to expand on this same document for the benefit of the Dowager Grand Duchess Christina. She was an important member of the European aristocracy, and Galileo was probably wise in trying to keep her on his side instead of letting the duchess' personal beliefs degrade into a hatred of modern astronomy.

Galileo beseeched the lady to understand that his interpretation of the universe did not go against the Bible. Instead, he argued, many people had misinterpreted the words of the Bible and therefore came to believe untruths that were in conflict with reality. In addition, Galileo wanted the duchess to agree with him that science and religion were not necessarily in constant connection with one another—a rather controversial point of view. His letter continued:

Before a physical proposition is condemned it must be shown to be not rigorously demonstrated-and this is to be done not by those who hold the proposition to be true, but by those who judge it to be false. This seems very reasonable and natural, for those who believe an argument to be false may much more easily find the fallacies in it than men who consider it to be true and conclusive.[32]

Galileo must have felt incredibly isolated at times such as these, when he was forced to insist that he was a faithful Catholic and that he believed fully in the truths of the Bible. Though perhaps the letters convinced some of European society of a necessary separation between the Bible and scientific research, the overall effect of his letters to Castelli and the Dowager Grand Duchess Christina was less positive. They served mostly to further demonize him in the eyes of the most devout members of the Catholic Church and the Inquisition. It was no help to his case that several members of the Inquisition who had been involved in the investigation and the subsequent execution of Giordano Bruno were working on the case against Galileo.

Many scientists of Galileo's era felt similarly boxed in by the authority of both the Catholic and the Protestant Churches of Europe, and yet they continued to believe in the existence of God as a creator. It seems likely Galileo felt the same way, mostly because he claimed as much time and time again, even in private letters. Despite his almost automatic approval of the words of the Bible, Galileo did not seem to have any modified religious beliefs, or at least not any that he wrote down for posterity. Unlike contemporaries who found relief in the new Protestant Churches sweeping across the continent, Galileo made no move to find solidarity with a group of less-critical churchgoers, which implies that he still found some solace in Catholicism.

Though Galileo was somewhat protected from backlash because of his personal relationship with Pope Paul V, he was not free from repercussions. In 1616, the same year in which Galileo was questioned by the Inquisition, the pope placed Copernicus' De revolutionibus orbium coelestium (On the Revolutions of the Heavenly Spheres) on the Index of

Forbidden Books, which was a list of publications deemed to be heretical in the eyes of the Catholic Church.[33] However, Galileo was still allowed to continue his studies in search of evidence and to use the geocentric model as a theoretical device.

Fortunately for Galileo, the Inquisition's investigation once more resulted in no formal charges nor any trial. The astronomer's growing file, however, was kept and was consequently updated over the course of the coming years.

Chapter 13: Discourse on the Tides

1616 was an incredibly busy year for Galileo, who not only visited the pope, negotiated terms with Cardinal Bellarmine, and suffered the investigations of the Catholic Inquisition; he also wrote a new book. His Discorso Sul Flusso E Il Reflusso Del Mare (Discourse on the Tides) was published that year, and it outlined theories on how oceanic tides were linked to heliocentrism. While looking up into the visible parts of our solar system, Galileo not only pondered the mysteries of the planets but how those heavenly forces might be affecting physical reality on Earth.

Despite the fact that the political environment of 17th-century Europe was quite firmly set against the idea of the Earth circling the Sun, Galileo could not help but pursue evidence to support the model of the universe he believed to be correct. Everything he recorded and observed seemed to prove Copernicus correct. Galileo considered the heart of the fact he wanted to prove—namely, that the Earth moved around the Sun—and wondered if that very planetary movement was responsible for the movement of the tides. His theory was further developed by his own observations at Venice's docks, where the collected water in ship's hulls would slosh forward and backward once the ship itself had come to a sudden stop. It was a phenomenon that he discussed at length with his colleague, Paolo Sarpi, a fellow Italian polymath of the day.

In Galileo's view, high and low tides could result from the forward and backward movement of the ocean's water when a body of water slows down or speeds up. As a staunch Copernican,

Galileo sought clarity on his theory of heliocentrism through this explanation of tidal movements. To this end, he determined that the tides were caused by the double movement of the Earth around the Sun and the spinning motion of the Earth. Due to the variable motion of the Earth, he posited, the seas either quickened or decelerated, much the same as the water in the merchant ships he observed alongside Sarpi.

Galileo, who was also focused on the understanding and development of pendulums during this time, found that the pendulum could help explain his burgeoning theory of the tides. Using a giant pendulum in his theoretical model of tidal movement, Galileo envisioned the Earth as the ball at the end of the instrument. In stacks and stacks of notes, Galileo illustrated this pendulum moving to and fro in respect to the theoretical movements of the Earth. He guessed that the Earth must swing back and forth every six hours to account for the related movements of the oceanic tides.

Galileo did not manage to uncover any indisputable clarification for the regular movement of high and low tides, especially considering the daily differences in the time at which high and low tides occur. Nevertheless, he continued trying to develop his theory for the rest of his life.

This was one scientific theory in which Galileo did not have the support of his friend, Johannes Kepler. Though his work was highly praised by Kepler, Galileo did not agree with him that the Moon was responsible for producing oceanic tides. Instead, Galileo believed that the tides were caused by the movement of the Earth around the Sun, as well as the rotation of the Earth upon its axis. It would eventually turn out that Kepler had this one right when Isaac Newton published Philosophiæ Naturalis Principia Mathematica (Mathematical Principles of Natural Philosophy) in 1687.

While Copernicus had appropriately seen that the planets spin around the Sun, it was Kepler who accurately characterized their circles. Utilizing the perceptions of Copernicus, which he had acquired from his previous employer Tycho Brahe, Kepler found that the trajectories of all known planets followed three laws, which became known as Kepler's laws of planetary motion.

These laws can be explained as followed:

1. All planets move about the Sun in curved circles called ellipses, having the Sun as one of the foci of the ellipse.

2. As a planet moves along its ellipse, the speed of the planet is not constant.

3. There is a precise mathematical relationship between a planet's distance from the Sun and the time it takes that planet to complete one revolution around the Sun.

Kepler's laws were years in the making since he required large amounts of data to change his own mind. Like numerous logicians of his period, Kepler began his career with a traditional Greek-style conviction that the circle was the most perfect route in the universe. He thought that since the planets had been created by God, their paths in space must be circular. For a long time, he attempted to use Brahe's objective facts of the movements of Mars to coordinate with a roundabout circle. Eventually, however,

Kepler was forced to admit that the data just didn't add up that way.

In the end, Kepler saw the truth and was convinced by his own research that it was not Earth's motion that caused the movement of the tides but the attractive forces of the Moon on the oceans. Galileo, unfortunately, remained quite fixated on the potential he saw in the movement of pendulums. In fact, though the pendulum could not satisfactorily explain the movements of the Earth, Galileo would eventually use them to improve upon other contemporary instruments.

Chapter 14: A Meeting with Pope Urban VIII

Pope Paul V died in late January of 1621 after suffering a series of strokes.[34] Following his death, there was the traditional election of his successor by the College of Cardinals, who chose Alessandro Ludovisi. Ludovisi chose to lead the Church as Pope Gregory XV, but his reign only lasted for two years. Upon the death of Pope Gregory XV in 1623, Maffeo Barberini was elected pope under the official name of Urban VIII.[35]

Barberini's ecclesiastical career had been upward bound from the beginning, and it was quite rapid considering his first major appointment had only been 22 years earlier, when he was made papal legate to Henry IV, King of France. Three years afterward, he had been given the title Archbishop of Nazareth and shot up to the post of papal nuncio to the French king. In 1606, he became the Cardinal of San Pietro (St. Peter) in Montorio.[36]

Urban VIII's main concern upon taking on the papacy was the missionary work of all Catholic denominations. He established a school for the training of missionaries and repealed a rule that only Jesuit missionaries could travel to China and Japan. It was important to the new pope that all recognized orders of Catholicism were able to participate in missionary work since he believed it was his holy mission to spread his particular faith across the face of the entire Earth.

Summarily, Urban VIII was not immediately concerned with overthrowing any of the decisions made by his various predecessors on scientific theory. He was aware that Pope Paul V had had an understanding with Galileo Galilei about appropriate publications and teaching methods, and he saw no reason to forego that agreement—especially because he and Galileo were already well acquainted and had been for more than a decade.

Maffeo Barberini had always been an accomplished man, as well as a socially mobile and well-educated member of society. When Galileo came to Florence in 1610, Barberini met him at a formal dinner with mutual acquaintances and found a deep respect for the scientist's intelligence as well as his sense of humor.[37] During that meal, Galileo heard sharp criticisms of his scientific perspective on the physical properties of floating objects, and Barberini encouraged Galileo's ideas against the other members of the clergy. From that point onward, the two men remained on friendly terms, with Barberini having no trouble understanding the differences between Galileo's science and his own realm in the Church.

The two met six times during Galileo's 1623 visit to Rome, during which time the new pope readily agreed that Galileo could write and publish on the subject of Copernicanism. His one stipulation, like that of Pope Paul V, was that it should only be presented as theory, not as hard truth. Upon seeing early versions of the manuscript upon which Galileo was working, Urban secondarily requested that the author include the geocentric theory alongside heliocentrism. This would have a massive outcome on the book, which would be completed and published in 1632.

For nearly a decade, Galileo enjoyed a warm relationship with Pope Urban VIII, and he probably felt the most at ease in his forward-thinking work than at any other point in his career. While Galileo breathed a little more easily, the pope set his sights on establishing his own authority as the highest within all of Italy. Unlike Pope Paul V, whose great aim had been to reunite all of Europe with Catholicism, Urban VIII instead chose to transform his own region into the most powerful realm on the continent.

The pope's goal was to acquire and coerce the smaller principalities of Italy into the papal state, which he planned to center on the city of Urbino. In 1626, the Duchy of Urbino was consolidated under his control, and soon afterward, Mantua became politically aligned with the papacy as well.[38] Much of Urban's plan involved political manipulation; therefore, he supported less wealthy rulers in regions he wanted to annex instead of supporting Europe's most elite families. In this way, Pope Urban VIII stealthily collected lands that otherwise would have gone to the likes of the Habsburgs or de' Medicis, Europe's most eminent and influential families.

In doing so, Urban VIII regularly supported Protestant families over Catholic leaders, all in the name of his broader goal. The pope went so far as to goad feuding regions into the Wars of Castro in 1641, after which he added Castro into the Papal States. Urban VIII was the last pope to broaden the ecclesiastical region, and he did so ruthlessly. Not bound solely by political and financial manipulations, Urban VIII likewise settled upon the grand task of stockpiling arms within the Vatican. He also set up an arms production line at Tivoli, about 40 kilometers (25 miles) east of Vatican City. To further furnish the armory, huge bronze supports were plundered from the ancient Roman

Pantheon, which inspired the people of the city to gossip terribly about the pope. The saying went, "Quod non fecerunt barbari, fecerunt Barberini": "What the brutes did not do, the Barberini did."[39]

Galileo, for his part, did not seem to have made any efforts to support the ongoing strategies of his influential friend. Of course, he would have had to become quite careful about which wealthy families he approached for scientific patronage, as aligning himself with one of the pope's many new enemies could put him in a very difficult situation. On the other hand, with Urban VIII himself making deals with Protestants, the new pope's reign must have simultaneously felt chaotic enough for some non-biblical, scientific logic to fall unnoticed through the cracks of the Inquisition.

Pope Urban VIII and his family gave generously to artisans in their cities on a stupendous scale. The pope personally exhausted his bank accounts to bring his favorite polymaths to Rome and in subsidizing different significant works, such as the stone carvings created by Gian Lorenzo Bernini. Bernini's work is considered the first of its kind in the world, and it encompasses a style called Baroque. Fascinated with Bernini's carvings, Urban VIII brought the artist to Rome to create sculptures for Rome's greatest residences and buildings, including the Palazzo Barberini and St. Peter's Basilica. Various individuals from Barberini's family likewise had their resemblances hewn in stone by Bernini, including, of course, the pope himself.

Sculptures were not the only artistic love of Urban VIII and his family members. They also poured funds into commissioning paintings by the likes of Nicolas Poussin and Claude Lorrain. One of the most self-aggrandizing artistic works paid for by the pope was the ceiling mural in the salon of the Palazzo Barberini. Painted by Pietro da Cortona, the intricate mural depicts various allegorical characters such as Time and Providence, seemingly working together to ensure the selection of the most righteous pope. The work was named the Allegory of Divine Providence and Barberini Power.

The result of all these splendid artworks, building projects, and military measures was an immense papal debt—and this did not go unnoticed by Urban VIII's contemporaries. In 1636, individuals from the Spanish section of the College of Cardinals were so outraged by the irresponsible spending of Pope Urban VIII that they actually put together a plan to have the pope imprisoned or murdered. Once they did that, the schemers believed they would be able to supplant him with Laudivio Zacchia.[40] When Urban VIII made a trip to Castel Gandolfo, the plotters met covertly and talked about approaches to propel their arrangement. Be that as it may, they were found out. Upon discovering the plan to have himself usurped, Urban VIII fled back to Rome and declared that every cardinal should leave Rome and return to their local churches, thus breaking up the conspirators.

With this near-disaster averted, Urban VIII did not slow his spending to try to win back his critics. Instead, he continued to fortify his lands and spend freely on artistic works, so much so that by 1640, he owed as much as 35 million scudi.[41]

Understandably, perhaps, due to these incredible debts and the public outcry against his leadership of the Church, Urban VIII's later years saw the fervent persecution of any purported slander of his office. Only a few years remained for Galileo in which his friendship with Urban VIII would continue to his benefit, but sadly, Galileo did not seem to have taken heed of the warning signs.

Chapter 15: The Assayer

Freshly home from his very successful meetings with Pope Urban VIII, Galileo published Il Saggiatore (The Assayer). The book came out in October of 1623 in response to the assertion of his colleague, Orazio Grassi, that comets were solid objects flying through space. Grassi's opinion on the matter was correct; however, in the early 17th century, it was difficult for astronomers to make a confident claim one way or another. Galileo believed that Grassi's notion of giant flying rocks in space was incorrect, and in The Assayer, he insisted that comets must instead be made of light, like the rays of the Sun, that travel at high speeds. He reasoned that there were two types of substances in the universe, the first of which can be physically experienced and the second of which is intangible. Comets, he went on, are of the intangible sort, made only of light and nothing else.

Il Saggiatore would be the last and most critical of Galileo's work on the topic of comets. Much of the content of the book was inspired by the arrival of three comets, all in the year 1618. The very next year, Orazio Grassi published an anonymous treatise on the subject of comets, and it was met with a great deal of criticism, much of which was mistakenly aimed at Galileo. Since the work's author was unnamed, contemporary astronomers asserted that Galileo had written the discourse and was the man behind the solid comet theory. Galileo, in no way supportive of Grassi's theories, was forced to take a defensive action that turned into Il Saggiatore.

Il Saggiatore was distributed in Rome under the protection of the Lincean Academy, and it was personally devoted to Pope Urban VIII. The cover sheet of the book depicts the family crest of the Barberinis, which includes three honey bees. Always sure to flout his papal patronage, Galileo did well to appease the powerful family in such a way as to encourage further protection.

In the pages of the book, Galileo evaluated the theories of Grassi and decided that they were incomplete. Taking Grassi's primary question about the idea of divine bodies into account, Galileo put forward a general, logical way to deal with the examination of comets. As usual, his discourse quietly touted the truth of the Copernican universe model. Moreover, Galileo posited that Grassi's solid-state comet theory was one-sided and that it did not use an appropriate scientific method.

In The Assayer, Galileo wrote:

I may be able to make my notion clearer by means of some examples. I move my hand first over a marble statue and then over a living man. To the effect flowing from my hand, this is the same with regard to both objects and my hand; it consists of the primary phenomena of motion and touch, for which we have no further names. But the live body which receives these operations feels different sensations according to the various places touched. When touched upon the soles of the feet, for example, or under the knee or armpit, it feels in addition to the common sensation of touch a sensation on which we have imposed a special name, "tickling." This sensation belongs to us and not to the hand.

Anyone would make a serious error if he said that the hand, in addition to the properties of moving and touching, possessed another faculty of "tickling," as if tickling were a phenomenon that resided in the hand that tickled. A piece of paper or a feather drawn lightly over any part of our bodies performs intrinsically the same operations of moving and touching, but by touching the eye, the nose, or the upper lip it excites in us an almost intolerable titillation, even though elsewhere it is scarcely felt. This titillation belongs entirely to us and not to the feather; if the live and sensitive body were removed it would remain no more than a mere word. I believe that no more solid an existence belongs to many qualities which we have come to attribute to physical bodies-tastes, odors, colors, and many more.[42]

Galileo confidently asserted that the theory of solid-state comets was heavily influenced by religion and was therefore dogmatic. He lamented the fact that his colleague, whoever it may have been, had not properly used mathematics to illustrate his point. Indeed, Galileo spoke of mathematics as if they were the language used by God to design and create the universe. The Catholic perspective on mathematics, as Galileo believed and as he hoped others would too, is that numerical statements are common, real, and true and that they can reveal God's work.

In Galileo's day, however, there were highly touted philosophers and even mathematicians who considered mathematics to be nothing more than a construct of the human mind. Under the assumption that all human languages and calculations were disconnected from truth and reality, even the most elegant equations were rendered insignificant. To Galileo, such philosophies were nonsense. He worked with numbers and physical observations every day, and he wholeheartedly believed that scientific thoughts uncovered amazing truths about physical reality.

The conviction that God made the universe in a systematic manner had motivated the ancient thinkers to pursue the very arithmetic, sciences, and innovations that carried Europe into the contemporary age. To back up his own theories concerning comets and to illustrate his point about keeping mathematics in all scientific discourse, Galileo explained that God should be seen as the establishment of all information, and that since science was an exceptional portrayal of God's creation, one should hope to discover, after understanding it, the undetectable things of God.

Johannes Kepler agreed with his friend about mathematics and their fundamental role in explaining God's creation. Kepler acknowledged he could celebrate God through his numerical investigations, and his logic notes were regularly blended with supplications and commendations to his Lord. Kepler accepted that there was organization in God's creation and that the more Catholics perceived the significance of creation, the more profound their love would be for their creator. As far as Kepler was concerned, the main point of all examinations of the outer world was to find the rules which had been intrinsically woven into it by God, and Kepler believed that the only way to uncover these rules was to employ the language of science. In his astronomical and physical research, Kepler just wanted to be able to interpret God's thoughts and actions.

Galileo not only heartily believed the same as his friend, but he hoped that all of society would come to understand mathematics and scientific evidence in the same way. He had been warned by a succession of popes against pushing a Copernican agenda, but Galileo could not help but hope that he could persuade the world to understand that mathematics was God's language and that discoveries made using such a language were the opposite of heretical.

Chapter 16: Dialogue Concerning the Two Chief World Systems

In 1632, Galileo published a book whose contents he had been working on for decades. It was the culmination of all of his research and calculations in support of the heliocentric model of the universe, and finally, he was prepared to release it publicly. Of course, no great work went to the printing press before one's patron had a preliminary look—and in Galileo's case, that meant checking in with Pope Urban VIII. Based on their prior agreement, Urban VIII allowed Galileo to publish the book as long as it presented counterarguments in support of geocentrism. Urban specifically asked that his own views be written into the book, and it was this demand that would cause a great deal of trouble later on.

The book was revised accordingly, and upon publishing, it was called the Dialogo sopra i due massimi sistemi del mondo, (more commonly known by its English translation, the Dialogue Concerning the Two Chief World Systems). The Dialogue would be Galileo's most influential—and most controversial—publication. Having already earned the permission of the previous pope to continue his studies in non-Aristotelian astronomy, Galileo was similarly allowed by the new pope, Urban VIII, to publish his idea of a non-geocentric universe if it was presented as a theory alongside the church-supported geocentric model. Much like a school in which both biological evolution and intelligent design are presented to students, Galileo was forced to offer both models of the universe as theories with equal potential.

The Dialogue was the outcome of those papal instructions, and it was a unique work in that it was not written solely for academics. Instead of reading like a scientific article or a textbook, the Dialogue was just that: a dialogue between two sets of characters. One set of characters in the story were proponents of the heliocentric model, and the others were in favor of a geocentric model. The basic tenets of both models were explained via conversations between the characters Galileo created. As for the pope's special request, technically speaking, Galileo did just what Urban VIII requested of him.

Unfortunately, when the Dialogue Concerning the Two Chief World Systems went public in 1632, it became clear that there was a strong subtext present in the scientist's book. The main character in favor of geocentrism (the voice of the pope) was called Simplicio—a perfectly acceptable Latin name. However, it was a name that was just as often used to make fun of a dimwitted person then as it would be today. Simplicio argued his points in favor of the Sun circling the Earth but was constantly berated by the other characters in the book for his inability to grasp that it was the Earth that, in fact, circled the Sun. Simplicio was ridiculed throughout the text, making it quite clear what the author's views were on the universe, despite Galileo having been clearly instructed not to show favor for one model over another. Clearly, the Dialogue's author believed Aristotelian- and Pythagorean-style sciences to be out of date.

Below is an excerpt from the book. In this passage, Simplicio tries to guilt his rival, the Copernican Salviati, into embracing the sciences of the ancient Greeks because of their shared love of mathematics. Instead, Salviati responds that Pythagoreans from long ago condemned the publication of true mathematics, believing that such knowledge was too complex for the common reader.

SIMPLICIO. It seems that you ridicule these reasons, and yet all of them are doctrines to the Pythagoreans, who attribute so much to numbers. You, who are a mathematician, and who believe many Pythagorean philosophical opinions, now seem to scorn their mysteries.

SALVIATI. That the Pythagoreans held the science of the human understanding and believed it to partake of divinity simply because it understood the nature of numbers, I know very well; nor am I far from being of the same opinion. But that these mysteries which caused Pythagoras and his sect to have such veneration for the science of numbers are the follies that abound in the sayings and Writings of the vulgar, I do not believe at all. Rather I know that, in order to prevent the things they admired from being exposed to the slander and scorn of the common people, the Pythagoreans condemned as sacrilegious the publication of the most hidden properties of numbers or of the incommensurable and irrational quantities which they investigated.

Simplicio may have been unable to understand the finer points of the Dialogue, but Pope Urban VIII was perfectly capable of seeing that he was being ridiculed by the man for whom he had made so many concessions. Urban VIII quickly banned the book from being sold in Rome and set the Catholic Inquisitors upon the manuscript to make a record of all the heretical transgressions found in its pages. The friendship between Galileo and his long-time friend and patron was well and truly over.

For his part, Galileo claimed to have simply followed the instructions that had been given to him—any interpretation of his Simplicio character as anything but a figurative character was wrong, he insisted. Unfortunately for the scientist, life was about to become quite difficult. The Catholic Inquisition finally had the support of the pope to investigate and put Galileo on trial, and he had also begun to suffer greatly from poor health.

In fact, that same year, Galileo's doctors attempted to intervene with the Catholic Church on his behalf. He was frail by that time, and with failing health and vision loss, his physicians rightly worried that being subjected to the notorious tortures of the Inquisition would kill him.

December 17, 1632

We, the undersigned physicians, certify that we have examined Signor Galileo Galilei, and find that his pulse intermits every three or four beats, from which we conclude that his vital powers are affected, and at his great age much weakened. To the above are to be ascribed frequent attacks of giddiness, hypochondriacal melancholy, weakness of the stomach, sleeplessness, and flying pains about the body, to which others can also testify. We have also observed a serious hernia with rupture of peritoneum. All these symptoms are worthy of notice, as under the least aggravation they might evidently become dangerous to life.

Vettorio de Rossi

Giovanni Ronconi

Pietro Cervieri[43]

Though neither the pope nor his inquisitors made a point of mentioning the note or thereby lessening any such planned torture or punishments, it seems quite possible that the doctors' intervention had a positive effect. After all, Giordano Bruno had been burned for similar charges just 33 years earlier; by contrast, Galileo's subsequent trial and punishments for his pursual of science seem quite mild.

Chapter 17: Trial and Imprisonment

In 1633, Galileo was officially charged with heresy for teaching the Copernican theory—a theory that opposed the purported biblical version of a fixed Earth at the center of the universe. Galileo presented his letter from Cardinal Bellarmine to the inquisitors, but it was of little use. Representatives of Pope Urban VIII pleaded with Galileo to simply recant his position on the matter and beg forgiveness, but he refused, time and time again.

Vincenso Firenzuelo was given the task of interviewing the accused multiple times, and during each of these interrogations, he tried every way he knew how to extract a confession and an apology from the astronomer. At least one of these meetings gave Firenzuelo hope, which he gleefully expressed in a letter to Cardinal Francesco Barberini.

I entered into discourse with Galileo yesterday afternoon, and after many and many arguments and rejoinders had passed between us, by God's grace, I attained my object, for I brought him to a full sense of his error, so that he clearly recognized that he had erred and had gone too far in his book. And to all this he gave expression in words of much feeling, like one who experienced great consolation in the recognition of his error, and he was also willing to confess it judicially. He requested, however, a little time in order to consider the form in which he might most fittingly make the confession, which as far as its substance is concerned, will, I hope, follow in the manner indicated.

Your Eminence's most humble and most obedient servant,

Fra Vinc. Da Firenzuelo

Rome, April 28, 1633[44]

Perhaps Galileo had some hope that his old friend, Urban VIII, would have a change of heart, or perhaps Galileo was simply tired of trying to hide the truth about his perspective on the universe. At any rate, the pope did not intervene on Galileo's behalf, and despite finally giving in to the increasing pressure to recant his statement that the Earth revolves around the Sun, the scientist was found guilty of heresy on June 22, 1633. He was sentenced to imprisonment.[45]

The very next day, the sentence was amended from imprisonment to house arrest, and Galileo was moved into the comfortable estate of Ascanio Piccolomini, the Archbishop of Siena, Tuscany. Though the convicted man was not allowed to leave the estate, he had private quarters and was treated like a guest instead of a prisoner. Eventually, having been a well-mannered house guest with Piccolomini, Galileo was allowed to return to his own home in Arcetri, Florence. However, he was not allowed to leave the grounds. The living arrangement in Arcetri was a pleasant space in which to reside, and it was very near to the nunnery where both of Galileo's daughters lived. Sadly, Maria Celeste, who had changed her name from Virginia after joining the convent, died in mid-1634.

At Arcetri, Galileo received some of the most eminent visitors in the realm, including Ferdinando II de' Medici and the painter Justus Sustermans, also known as Giusto Sustermans. Sustermans took the opportunity during his visit to Arcetri in 1636 to paint a portrait of Europe's most famous scientist. His portrait, created in the new

Baroque style, is a realistic portrayal of a greyed, long-bearded man in conservative black with a thick white collar. Galileo's eyes seem to focus directly on the artist, and though he looks experienced and wise, he also appears tired and a little sad. Sustermans' portrait of Galileo has been considered by many to be the best likeness of the man himself.

Despite his heart condition, grief at the loss of his daughter, and ever-encroaching blindness, Galileo could not help but continue his work. Galileo's theories on the forces of gravity became an important part of his final publication in 1638, Discourses and Mathematical Demonstrations Relating to Two New Sciences. The book was one of the first treatises on what was indeed a new science: physics. Galileo's work on the subjects of attractive forces and interactions between physical objects would be one of the most important physics books of the century and the precursor to Isaac Newton's Principia in 1687.

In the pages of Two New Sciences, Galileo finally returned to the problem that had haunted him: the thickness of the ceiling of Hell. It had been fifty years since he had faced the Florentine Academy and supported the hypothesis of Antonio Manetti against Alessandro Vellutello, and it had been almost as many years since he had kept a terrible secret from the public, that his measurements concerning the dimensions of Dante's vision of Hell were incorrect.

Galileo had originally assumed that when the size of a dome was increased, its thickness would increase at the same rate. Therefore, when the length and width of the dome were doubled, he calculated the essential increase of its thickness at the same rate: Double the length and width, double the thickness. This was not, unfortunately, the correct relationship between the length, width, and thickness of a domed roof, and Galileo had figured out his mistake soon afterward. The reasons for hiding his mistake were fairly obvious, given the prestige, job, money, and acclaim Galileo's work on the project had earned him.

Nevertheless, the 73-year-old scientist found it was time to unburden himself and to bestow the true mathematical equation by which one could calculate the various features of a solid, strong, domed roof. In reality, roofs such as that described by Dante as covering the immense roof of Hell must increase in thickness much more quickly than they increase in length and width. Galileo expressed this relationship in what became known as the square-cube law. In basic terms, the square-cube law shows that as an object increases in size, its volume increases more quickly than its surface area.

This general law had already been proven by Johannes Kepler. Kepler had published a book called Nova stereometria doliorum vinariorum (New Solid Geometry of Wine Barrels) in 1615 that proved that as the dimensions of a barrel grew, so too did its volume. His findings came after furious research intended to prove that he had overpaid the wine supplier at his own wedding. He probably had, but not by as much as he had hoped.

Galileo's square-cube law is still used by engineers and scientists today as it has many more applications than it seems at first glance. The law explains not only that roofs and beams in buildings must become thick as their length increases, but it also explains why the bones of large animals, such as elephants, are much thicker than those found in smaller animals. For such animals to support their heavy weight, their bones would have to be incredibly thick.

As for Dante's roof of Hell, there was more to the issue than miscalculating the thickness of the roof, and that is probably why Galileo kept his secret for so long. Having used his perfected square-cube law to recalculate the dimensions of such a roof, he realized that the thickness of Hell's roof would have to be so thick to support its continent-sized length and width that there would be no space under it to house Dante's nine large levels of the underworld. There was simply not enough space under the surface of the Earth to hold all the millions of souls that had lived and died.

Perhaps Galileo would have confessed his mistake much sooner if it were only a case of numbers; he was clearly not prepared to take on responsibility for the apparent fact that Dante's vision of Hell was simply incorrect. Dante had referenced his poetic representation of the nine layers of Hell very carefully, drawing information directly from the Bible—and that meant that if Galileo were to come forward with his findings that no such place could exist as described, he would undermine some of the fundamental doctrines of the Catholic Church.

Given that he was already under house arrest, it probably didn't seem like the kind of secret worth bringing to his grave. So, he endeavored to have the new book published. Given the papal ban of all books written by Galileo, the eminent scientist was reluctant to publish Two New Sciences in nearby Venice, even though he had an offer. He had a hard time finding a publisher outside of Italy, though; he could not find anyone willing to publish in Germany, France, or Poland. It was not until the manuscript was presented to Lodewijk Elzevir of South Holland that Galileo's work finally hit the press. The reach of the Catholic Inquisition was somewhat less lethal in that part of Europe, which explains Elzevir's willingness to add Two New Sciences to the tomes of his family's great publishing house.

Already in the custody of the Church—albeit in reasonable comfort in his own home—Galileo had little to lose when Two New Sciences began selling out in cities across Europe. Copies of the book soon appeared as near as Rome, but evidently, the Inquisition saw little reason to bring their prisoner back to trial. It had already been established that the pope had no desire to torture or kill Galileo, and therefore, no further legal measures were taken.

A general preface and a

From the 17th century to the present day, many have promised Galileo the breadth of modern science. His discoveries are numerous, and he was the first to see the Moon Mountains and Jupiter's moons using the telescope. It also defined the shape of the trajectory of the projectile objects, and established the free fall law based on several experiments. He was also known for defending and promoting the Copernicus theory, using the telescope to look at the sky, by inventing a microscope, and by throwing stones from towers and masts. And by treating pendulum and watches. And also by defending the relativism of the movement,

And his innovation of mathematical physics. And also being the first "real" experimental world. The trial of Galileo by the Catholic Church and its subsequent presentation of the image of the modern man with the heroic role in the history of the "conflict" between science and religion were perhaps the most important factors in his fame. These are the achievements of a 17th-century Italian man, a man who was the son of a court musician, and left the

University of Pisa without obtaining a graduation certificate. And it's not an easy achievement at all.

One of the good things that distinguish important and exciting times, such as galileo's, is the abundance of interpretation, and the multiplicity of interpretations. Since Galileo's death in 1642, his work has been subject to various interpretations and is the subject of endless conflict. Begging for Galileo and his works is a fact to create an amazing historical story (Segre 1991, Palmerino and Thijssen 2004, Finocchiaro 2005), but that is not the purpose here.

On the philosophical side, many writers used Galileo to embody their various subjects, usually when talking about the scientific revolution or the nature of useful science. Whatever modern science or science in general was described, its launch was with Galileo. He knew the early 20th-century heritage around Galileo divided his works into three or four sections. His work (1) in physics, (2) in astronomy, (3) in the curriculum, and this last section may be attached to Galileo's approach to the interpretation of the Gospel, and his views on the nature of the evidence or the evidence. In this heritage, we find traditional treatments of Galileo's physical and astronomical discoveries and their relationship to the views of those who have been staging. While many may ask more important questions philosophically, such as: How does Galileo's mathematics relate to his natural philosophy?

And how did he make the telescope and then use the results he got to demonstrate the Copernicus theory? (Reeves 2008). Was Galileo experimental? (Settle 1961, 196, 1983, 1992; Palmieri 2008) Was he platonic in terms of mathematics? (Koyré 1939). Was he a aristotelian who emphasized the yeast? (Geymonat 1954). Was al-Bashir in modern positius science? Drake 1978, or maybe he was a licensed man? (Machamer 1998).

Did he really use a refined skultis approach to pampering? (Wallace 1992). Or was it a no-man's sake, but a shaved shaver in the sky as geniuses shave? (Feyerabend 1975). Behind all these allegations lies a quest to put Galileo in an intellectual context, highlighting his achievements. Some focus on Galileo's use of practical, traditional, traditional/ artistic/ engineering (Rossi 1962). Some for his sports productions (Giusti 1993, Peterson 2011,, Feldhay 1998, Palmieri 2001, 2003, Renn 2002, Palmerino 2015)."

Others on his mixed (or dependent) mathematics (Machamer 1978, 1998, Lennox 1986, Wallace 1992). Some also focus on Galileo's use of atomic philosophy (Shea 1972, Redondi 1983) and his use of the theory of momentum from the Hellenic and Medieval periods (Duhem 1954, Claggett 1966, Shapere 1974). Also on the idea that discoveries bring new data to science (Wootton 2015).

But almost everyone in this heritage seems to have viewed each of galileo's disciplines, i.e. physics, astronomy, and the curriculum, as if in a way independent of itself, embodying different directions of Galileo. Contemporary historical research follows the latest intellectual trends and focuses on studies that add new dimensions to our understanding of Galileo.

By studying his eloquence (Moss 1993, Feldhay 1998, Spranzi 2004), he built power in the social environment in which he lived (Biagioli 1993, 2006), as well as his personal quest for recognition and recognition (Shea and Artigas 2003). In general, these studies take care of various aspects of social and cultural history, namely royal and papal culture, for the period in which Galileo was active (Redondi 1983, Biagioli 1993, 2006, Heilbron 2010).

In a way that may seem an intellectual setback, we will summarize galileo's research in physics and astronomy. We will also creatively explain how all of these researches have been adapted to a unique research. During our biography, we will explain why Galileo felt at the end of his life the need to write his book "Articles on Two New Sciences", which is truly a complement to his total project and not just old research to which he returned after he lost his sight, became trapped in his home,

After his trial. In particular, we will try to explain why these two sciences, especially the first, were very important (a subject that has only recently been launched, such as Biener 2004 and Raphael 2011). We will also introduce Galileo's approach and mathematics (and we refer the reader here to some recent work on this, such as Palmieri 2001, 2003). In conclusion, we will briefly address Galileo, the Catholic Church and his trial.

Galileo's science story

The philosophical course governing Galileo's intellectual life is his increasing lyrical desire to renew the concepts of natural philosophy and the method of its practice. This was true after he left Padua in 1611 and moved to the Medici Palace in Florence, asking the Duke to call him a "philosopher" alongside the résumé. His request was not only an expression of his desire to strengthen his position in the palace, but also his far-reaching objectives.

What Galileo accomplished in his life was to replace the set of traditional analytical concepts associated with the Aristotelian heritage of natural philosophy (Aristotelian essays), a controlled replacement, and the introduction of alternative mechanical concepts, which he received most of the subsequent development of the "new science", which in one way or another became the name of "new philosophy". In fact, Galileo's approach to "scientific revolution" was the approach. (Yes, there was a "revolution" of this kind, with full respect for Shapin 1996 and others, see some anthology in Lindberg 1990, Osler 2000

In describing galileo's achievements, a number of scientists used some psychological terminology. Such as :"Opening new mental models" (Palmieri 2003), or "A New Model of Understanding" (Machamer 1998, Adams et al. 2017). Whatever that was, Galileo's main objective was to destabilise the Aristotelian physical essays about the celestial element (ether or fifth element), about the four earth elements (dust, fire, water, air), and about their varying movement directions (circular, straight up and down).

Replace all of this - except the material-included - with new ways to describe the properties and movement of the material using the mathematics of balance and proportionality (Palmieri 200) - the simple Archimedes machines, the scale, the oblique surface, and the wind, to which Galileo added the pendulum (Machamer 1998, Machamer and Hepburn 2004, Palmieri 2008). By doing so, Galileo changed this usual way of talking about the material and its movement, and therefore authorized the mechanical tradition, which would determine the characteristics of modern science later to this day, but this is a subject that is long overstated (Dijksterhuis 1950, Machamer et al. 2000, Gaukroger 2009).

In the context of galileo's achievements, it is useful to note his interest in creating a mathematical theory that is unique to the material that makes up the entire universe. He did not realize that doing so was his ultimate goal until he actually began working on his book "Essays on Two New Sciences" in 1638. Although since 1590 he began to work on some problems of the nature of the material, he could not complete his author before 1638. In particular, not before the publication of the "Messenger of the Stars" in 1610, nor before 1632, when he published a dialogue on the two main systems of the universe. Before 1632, he had neither the theory nor the sufficient evidence to support his claims about the single material. Although he had long thought about the nature of the material and tried to find the best way to describe it before 1610, the idea of the theory of the monotheism should have waited until the author had finished the principles of the movement of the material, which did not happen before the publication of the book "Dialogue on the Two Main Systems of the Universe".

Galileo began criticizing Aristotle in 1590, in the manuscript "In Motion". In the first part of this manuscript, Galileo directed the inspiration of Aristotle's theory of earthly matter. Aristotle believes that the earthly material is formed by four elements (dirt, air, water, and fire), including heavy ones, including light. Its movement up and down varies depending on its weight or lightness. Based on Archimedes' principles on floating objects and balances, Galileo claimed that there was only one principle of motion, where he believed that the weight (or gravity) was the cause of all natural earth movements.

The light (or levitas) can be explained by the fact that heavy objects move to remove other parts of the material in a direction that explains why the other parts are rising. Galileo had trouble explaining the nature of gravity (or gravitas) but claimed in the manuscript "In Motion" that the moving scale arm model could be used to solve all movement problems.

Galileo soon noticed that what he suggested was not enough, which led him to begin research into the fact that the weight varies depending on the specific gravity of objects of the same size. Galileo tried to come up with a concept of weight that would be specific to all materials. However, he did not succeed in doing so, and this is probably why he did not publish the author of "In The Movement". There seemed to be no way to find standard measures of the content that were common to different materials. So to this point, Galileo had no better alternatives.

A period later, in 1600, Galileo presented the concept of momento, a concept similar to that of power, which affects the body in a momentary, somehow proportional to weight or gravity(Galluzzi 1979). But yet Galileo had no way of measuring or comparing the values of the nucleus of different bodies,

In his dedications in this early 17th century, he tried to find a way to subject all materials to a single proportional scale. Galileo tried to study acceleration on the sloped surface and understand the consequences of its change. In this regard, Galileo attempted to detect the effects of the impact of objects of different gravity, or why they were subjected to varying effects. But the solution to the weight and movement problem remained out of reach.

One of the problems created by the simple archimedes machines, particularly the libra, which Galileo used as a model to illustrate his ideas, is that it is difficult to visualize their work in a dynamic way (but see Machamer and Woody 1994). In all these simple machines, except for the sloped surface, time is not usually the main concern. When we observe a scale, we don't usually care about the rate at which one of his palms are lower or the height of the body placed on the other (but Galileo does so in "Comments to Rocco" approximately 1634-1645,

See Palmieri 2005). On the contrary, it is difficult to illustrate a dynamic phenomenon such as the rate at which the speed of different objects changes using the balance model, one of which rises and the other decreases due to the different weights placed on them. Thus, the dilemma of how to describe time and the time of the impact of objects, which have long puzzled Galileo, remained unresolved. During his lifetime, Galileo could not find organized relationships between the nucleus of gravity, the rise of the fall, and the power of shock. But on the fifth day of the "dialogues", he became a concept of shock force, which, after Galileo, would become one of the most effective ways of thinking about the material.

Galileo worked long between 1603-1609 on numerous experiments on the sloped surface and, most importantly, on the pendulum. Through his work on the pendulum, Galileo once again saw how important variables such as acceleration, and of course time.

The phenomenon of synchronization — equal to time when the length of the rope is equal even with different weights — also saw the possibility of using time to describe the balance (or ratio) to be highlighted when representing the movement. The possibility of taking time is also a major variable rather than weight. He also became increasingly convinced of the importance of acceleration and time after working on the power of shock and sloping surfaces, and in 1608 he developed a small research on acceleration, but it was never published.

We conclude from the above that Galileo did not reach the law of free fall only after long suffering to find the right sayings for the science of matter and the new movement. Galileo, perhaps since he drafted the book Mechanics in 1594, acknowledges that natural motion can accelerate. However, the idea of comparing the accelerated movement with time came to mind only later, especially after it failed to find any connection between it and the place or the natural attraction.

He may have observed that, by nature, the speed of objects increases as they move downwards. Especially in the pendulum, in the free fall on sloping surfaces, and during the movement of the projectile. Also during that period, Galileo began researching the force of shock, the strength that the body gains during its movement resulting from exposure. Galileo has long believed that the correct theory about these changes should describe the change of objects depending on their location on their paths, specifically the height seemed very important. As the force of impact is directly related to altitude, it also shows that the movement of the pendulum is essentially balanced about the height of gravity (and also time, although the synchronization has been somewhat important).

Galileo discovered the law of free fall, expressed in the box of time, during his experiments on sloping surfaces (Drake 1999, v. 2). But he tried to explain this relationship, and the relationship of the proportional medium, by the relationship between speed and distance. Galileo's subsequent and correct definition of natural acceleration was time-dependent,

After realizing the physical importance of machamer and Hepburn 2004, to see a different analysis of Galileo's discovery of free fall law, see. Renn et al. 2004). Galileo, however, did not publish any of his ideas on the importance of time in the movement until 1638, when he published His Articles on Two New Sciences (Galileo 1638/1954). But let's get back to our original theme.

Galileo began working on the telescope in 1609, and the following year he published his book The Messenger of Stars, which contained his first telescopic discoveries. Galileo's time on the telescope is seen by many as a period of recreation away from his work in physics.

Galileo's findings can, of course, be described in many ways, but in keeping with our purpose here, we say that its results are truly exciting, because they marked the beginning of a departure from the framework of the separation of celestial and terrestrial matter (Feyerabend 1975).

Perhaps the clearest witness to this is Galileo's comparison of the Moon

Mountains to the Bohemia Mountains. The abandonment of this binary distinction between heaven and earth implied that matter was the same, whether terrestrial or heavenly. If so, there must be one natural movement as well. Thus, we conclude from the above that there must be one law of movement that applies to the earth, heaven, and even hell. This is a far greater claim than Galileo claimed before in 1590.

In addition, he describes the discovery of the four moons surrounding Venus, which he called , for political purposes, the Medchi stars (after his patrons, the Medchi, the ruling family of Florence at the time). One of the problems of Copernicus's theory was that the Earth had a moon orbiting it, which made it unique. But with the discovery of Jupiter's moons, that problem has been solved, and the Earth has been reestablished like the rest of the planets. For an amazing account, more detailed information about this period of Galileo's life, and its motives, seen recently published (Biagioli 2006, Reeves 2008, and articles in Hessler and De Simone 2013).

In 1611, at the request of Cardinal Robert Bellarmine, the professors of the Roman College of The Roman College recognized Galileo's telescopic discoveries. Father Calvos had opposed its confirmation, because he thought the moon's surface was not different. But he changed his mind shortly afterwards.

A few years later, in 1612, galileo in "Letters on Solar Spots" provided other reasons for ending the separation between celestial and terrestrial matter. In summary, the sun appeared to have maculae spots, that it was moving in a circular motion, and more importantly, the planet of the planet had a moon-like fold.

This latest discovery helped greatly to determine the physical location of the space between the Sun and the Earth, as well as the fact that the planet is orbiting the sun. In these speeches, Galileo also claimed that the new telescopic evidence supported copernicus theory, as the presence of phases of the sorority strongly opposed the ptolemaic arrangement of planets.

Later in 1623, Galileo presented a completely erroneous thesis on the article in his book Analyst. He tried to prove that comets were terrestrial phenomena, so their properties could be interpreted using photorefraction. While this book is considered a verse in scientific writing, it is very strange that

Galileo argues against the super-lunar nature of comets. This theory was earlier presented by the great Danish astronomer Tikhon Brahe.

Even with all these changes, Galileo had to work on two things. The first is to find some general principles about the nature of the movement according to his theory of the material. Specifically, given his peddle of Copernicus theory, he had to find a way to think about the movement of matter on the moving Earth. For Galileo, it wasn't just a transition from Ptolemy's Earth's central model to the sun's central model, as at Copernicus, !.!.!

It was also a transition from a mathematical model of planets to a deliberate physical description of the universe. Galileo also had to describe the planets and the Earth as real physical objects. In this, Galileo differed greatly from Ptolemy, Copernicus, and Tycho Brahe, who departed from this class with his theory of the celestial nature of comets, and by flirting with the physical model (Westman 1976). So in Galileo's new vision there is only one type of matter, and this material has one kind of natural movement as well. Galileo therefore had to invent (or discover) the principles of spatial motion suitable for a central sun, around which the planets orbit, and for a earth that rotates every day.

Galileo did so by introducing two new principles. On the first day of a "dialogue on the two main systems of the universe", Galileo claimed that all natural movements were circular. On the second day of the dialogues, he presented his version of the famous principle of relativity of the conscious movement.

The latter holds in its folds that common movements cannot be made, but only different movements can be perceived. Combining these two principles together means that the material shares a kind of movement, the circular motion, and any movement other than this common movement, for example, the movement up or down, can be realized directly. Galileo certainly wasn't the first to bring these principles, they were taken to. The difference, however, is that no one considered these two principles necessary for galileo's reasons, particularly with regard to the theory of the unique cosmic material he came up with.

On the third day of the dialogues, Galileo dramatically supported the Copernican theory. He made Salviati, Galileo's character, draw the attention of Simplisio, the confused Aristocratic character, to the new telescopic surveys, particularly those showing that Jupiter has folds, and those that show that Jupiter and Mercury are not far from the sun. This is to convince him of a chart of planetary positions that is never different from the Copernicus model.

On the first day of these dialogues, Galileo reiterated his claims in his book The Messenger of The Stars that the Earth should be like the moon: spherical, dense, solid, and contains rugged mountains. It certainly refrains from the moon being a celestial area, as some Aristotle thought.

In any case, the topics of the dialogues are more complex than we have alluded to here. It is true that Galileo has adopted the theory of circular nature movement, which involves the fact that all things in the Earth, and in the atmosphere, are in common with the Earth, and for this reason the principle of observed relativity of motion applies to phenomena such as the dropping of a ball from the moving ships, but in other places it offers a straight natural movement.

For example, on the third day of his dialogues, Galileo provides a partial explanation of the Coriolis effect on the winds surrounding the Earth, using straight motion (Hooper 1998). Also, when he presented his evidence of copernicus theory by explaining how the three-way moving Earth mechanically affected the movement of tides, he distinguished his theory of matter from the water ratio to the ability to retain the thrust of the violent movement of the basins, in order to generate reverse motion. This wasn't, of course, the first time Galileo had water

He had previously discussed submerged objects in his 1590 book In Motion, but more importantly, he developed a lot during his previous debate on floating objects ("Article on Floating Objects", 1612). In fact, much of this controversy has sparked research into the exact nature of water as a substance, and the appropriate type of mathematical proportionality to describe water and objects moving on it (see Palmieri, 1998, 2004a).

The final chapter of Galileo's scientific story begins with Galileo's publication of "Essays on Two New Sciences" in 1638. The second flag, which has been discussed (as it is said) in the last two days, deals with the principles of spatial movement, which have received a lot of attention in galileo's heritage. Here we find Galileo's disclosure of the law of free fall, the trajectory of the projectile in the form of parabola, as well as his physical "discoveries" (Drake 1999, v. 2). However, the first science, which was presented in the first two days, was often misunderstood, and little was discussed. This science has traditionally been mischaracterized and misleadingly called the science of material mechanics. Which made him seem fit to attach him to the history of engineering,

This science is still being studied today. But the first science was not actually material mechanics, but an attempt by Galileo to present mathematical science to his theory of the material unit (see Machamer 1998, Machamer and Hepburn 2004, and biener 2004 for a detailed and extensive account). Galileo noted that, before he could formulate a science about the movement of the material, he should try to clarify the possibility of describing the nature of the material as a matter of course. I believe that the sporty nature of matter and the mathematical principles of the movement belong to the mechanics. It is the name galileo gave to the new way of philosophy. But let us recall that the spiritual appeal has not been successful.

So on the first day, Galileo begins to discuss how the causes of the collapse of the greenhouses can be described mathematically (or geometrically) as the causes of the collapse of the greenhouses. It is a mathematical description of the basic nature of the material. It also continues to explain its different characteristics, but avoids looking at some issues that may be based on the infinite number of atoms. Such as the question of the structure of the material, its characteristics depending on its weight, the characteristics of the medium in which the objects move, and the reason why the body remains a physical unit. But the most famous of these discussions is that Galileo addressed the acceleration of falling objects and how they will all fall just as quickly in the void even with their different weights.

Then on the second day Galileo presents the mathematical foundations of body breakage by addressing the problems of the substance as similar in a way that work in balance and compassion. This is what he started in 1590 — although he believes here that he has succeeded in his quest this time — as well as explaining how parts of the material are grouped together and hardened, all by explaining how they are separated into pieces. However, Galileo did not succeed in reaching a definitive explanation for the aggregation of parts of the material, perfirst because he felt he had to deal with the tiny quantities in order to really solve the problem.

The second flag, covering the third and fourth days of discorsi dialogues, addressed the appropriate principles of spatial movement. But the movement of the whole material this time, not just the movement of earth objects, is also one of the distinctive things here, putting time sayings, and accelerating as the basis for action. It should be noted here that Galileo returned to, or felt, the need to include some non-arbited views on the movement, which he had previously worked on in 1590. The most famous example of this is this "creative intellectual experience", in which two objects of the same material are compared but in different sizes, and indicates that, according to Aristotle, these two bodies will fall at different speeds, and the heaviest will fall more quickly. So when the two bodies are combined, the lightness of the smaller body will reduce the speed of the larger body and therefore the combined bodies will fall less quickly than the heavier body alone, as in the first case. But, here's the witness, it can also be said that when the two bodies are combined, !.!.!.

And counting them one huge body, the speed at which this body falls will be greater. Thus, he concludes that there is a clear contradiction in the Aristotelian vision (Palmieri 2005). It is then devoted to the fifth day to address its total principle about the power of the material during its movement resulting from its exposure to shock. This principle, called Galileo, deals with the power of shock, two objects affecting each other. Galileo here offers no solution to the problem, which will only be resolved later with René Descartes, who, according to Isaac Beckman, may have turned it into a problem of finding points of balance of colliding objects.

This brief tour of Galileo's work serves as an introduction to understanding the change he has achieved. He worked on a new science of matter, a new physical description of the universe, and a new science of spatial movement. All this using a mathematical method based, albeit somewhat differently, on the Euclidean proportional geometry, book 6, and on the works of Archimedes.

In this way, Galileo, therefore, raised the new mechanical science, the science of matter and movement, and renewed his sayings. Taking advantage of some of the basics of mechanical heritage, adding to it the saying of time, stressing the importance of acceleration. In the course of all this, however, he did not want to discuss the details of the nature of the article, in order to come to show that it was unified and to address it in such a way that the coherent discussion of the principles of the movement was possible. Thus, thanks to Galileo, the theory of the unity of matter became acceptable, and research into the nature of matter became one of the most important problems of the next new science of its time. The article was therefore truly significant.

www.ingramcontent.com/pod-product-compliance
Lightning Source LLC
Chambersburg PA
CBHW050403120526
44590CB00015B/1808